D0908339

RENEWALS ... 574

WITHDRAWN
UTSA LIBRARIES

EMPLOYEE THEFT INVESTIGATION

EMPLOYEE THEFT INVESTIGATION

J. Kirk Barefoot, C. P. P.

BUTTERWORTHS
Boston • London

LIBRARY
The University of Texas
At San Antonio

Copyright © 1979 by Butterworth Publishers

All rights reserved. No part of this book may be used or reproduced in any manner without written permission except in the case of brief quotations embodied in critical articles and reviews.

Butterworth Publishers
10 Tower Office Park
Woburn, MA 01801

Printed in the United States of America

Library of Congress Cataloging in Publication Data

Barefoot, J Kirk.
Employee theft investigation.

Bibliography: p.
Includes index.
1. Employee theft--Investigation. I. Title.
HF5549.5.E43B37 364.12 79-9764
ISBN 0-913708-33-X

*To Frank A. Seckler,
who would have enjoyed
seeing his concepts
developed in this form.*

Acknowledgments

A special acknowledgment must be given for the many valuable suggestions made by Mr. George H. Harris, Director of Security, Ace Hardware Co. In addition to constructive criticism on the manuscript generally, Mr. Harris was of particular help with those sections pertaining to retail security activities.

Also, particular thanks go to Mr. John M. Haun, Jr., Manager of Corporate Auditing, Cluett, Peabody & Co., Inc. Mr. Haun gave valuable comments and help on the accounting portions of Chapter 1.

I wish to recognize the help of my wife, Cathie, who typed the first draft of the manuscript, and also the encouragement received from various members of the Corporate Security staff of Cluett, Peabody & Co., Inc., which helped sustain motivation during the many months of work on this text.

Foreword

Employee Theft Investigation is a book that will have versatile appeal. It is recommended as a significant reference for senior business managers in order that they understand the impact of internal losses and, more importantly, that they understand how to control those losses and thus become more competitive in the marketplace.

For the academic community, *Employee Theft Investigation* represents a major contribution to a field of study still in its infancy that thirsts for knowledgeable information concerning the complex problems of internal theft.

The third audience that will probably applaud the loudest after reading *Employee Theft Investigation* will be professional security practitioners. They will soon recognize that the author has paid his dues many times over. He has been on the firing line and in the trenches through every phase of the most complex internal business investigation. The methods and procedures explained in this text are from real-life cases. Moreover, throughout the past 25 years, the author has lectured at various seminars to thousands of security practitioners on specific aspects of the material in this text. He has received documented feedback from the most cynical of security executives that his techniques work—and in the business world we are constantly and rightfully being measured for results.

Most experts concede that the greatest dollar loss to business emanates from theft by the company's own employees, including the range of job titles from president to porter and crossing all socio-economic, sexual, racial and national origin barriers. Other major sources of loss include errors, shoplifting, burglaries, robberies, and various acts of God, but internal theft has the greatest impact on the net income of business. Therefore, it is fitting that *Employee Theft Investigation* has been written at a time when so many are earnestly seeking effective solutions to the insidious problem of internal theft. This is not a "How to Prevent . . . ," nor is it a theoretical treatise, but rather it is a practical text, written by one of the most respected security practitioners, which should be classified as "must" reading.

Employee Theft Investigation offers a sophisticated, pragmatic and credible resource for the senior business executive, the academic community, and the security practitioner.

Richard D. Paterson, CPP
Director of Security
Foremost-McKesson, Inc.
San Francisco, California

Contents

INTRODUCTION

Internal Theft: Introduction

Covington County, Alabama . . . . On Thursday, September 16, 1971, two men arrived by automobile at the Air Host Inn, adjacent to the Atlanta Airport in Atlanta, Georgia. After making proper inquiry at the desk, they proceeded to a guest room on the third floor where they met with a purported businessman who identified himself as "Doc" Long.

One of the two visitors was carrying a .44-caliber magnum Frontier Special tucked inside his waistband, concealed by a sport shirt. He was introduced to the occupant of the room by his companion and then proceeded to negotiate with "Doc" Long for the sale of ten thousand shirts. After haggling over prices that ranged from eighty-five cents to one dollar per shirt and discussing such trivia as whether the labels would remain intact or would be cut out, buyer and seller agreed on a price of one dollar per shirt (with labels).

An initial order of one thousand shirts and an "open to buy" on the remaining nine thousand was placed. Seven "samples" had been interjected into the meeting by the pistol-packing "salesman" as evidence of what the buyer might expect on receipt of his goods. The whole transaction consumed no more than thirty minutes.

At a price of one dollar per shirt, the conspirators were talking of a commodity which wholesaled at the time of the discussion for approximately four dollars and fifty cents and retailed for approximately nine dollars. This transaction was but the tip of an iceberg, part of a number of conspiracies and large-scale thefts which cost an Alabama shirt manufacturing company approximately one hundred thousand dollars in inventory losses for the year 1971.

In addition to the gun-toting "salesman," the other two participants in the meeting were the company security agent working undercover along with another company security agent posing as the buyer of the shirts. There were no police, federal or otherwise, involved in the case at this point, nor need there have been. Yet, the case would ultimately result in approximately eight pleas or findings of guilty on federal conspiracy charges. Virtually all the participants drew penitentiary terms.

The "Theft Tax"

The above case is similar in many respects to thousands which occur all over the United States each year: the gigantic rip-off of American business and industry which is done almost totally to the detriment of the American public. There is an old adage which says that nothing is inevitable except death and taxes. Whoever authored that phrase years ago undoubtedly had in mind taxes which were necessary to support the workings of government, education, and other public services. Without question, the pundit could not have foreseen a new tax which affects all of us today—namely the "theft tax."[1]

One authority states that for the year ending 1974, crime was costing United States businesses approximately five billion

---

[1]Barefoot, J. Kirk, *The Polygraph Story*, American Polygraph Association, revised third printing, October, 1974.

dollars per year.[2] The same authority estimated that two billion of this was due to embezzlement, fraud, and other inside thefts. Other authorities have estimated the figure on inside or internal crimes against U.S. businesses to be closer to three or four billion dollars per year. The results of this catastrophic disclosure take two forms: either business bankruptcy, or an increase in consumer prices which has been termed the "theft tax."

It must be recognized that all of the above figures are only estimated. Security authorities are quick to point out that these estimates are conservative in nature; if anything, the actual losses to U.S. businesses from internal crime could easily double the above figures. The truth of the matter is that there is no easy way to develop meaningful statistics for internal business crime. Most internal thefts are never reported to the police, and only a small percentage are ever reported to insurance carriers who normally provide fidelity bonds for the business community.

According to the Small Business Administration of the United States Department of Commerce, business bankruptcies doubled from 15,000 to 30,000 between 1970 and 1976.[3] In follow-up to this, the Small Business Administration has been quoted in the past as stating that from thirty to fifty per cent of business failures have been attributed to employee dishonesty.

Many larger companies, because of their size and their competitive positions in the marketplace, are able to absorb internal theft losses. At the beginning of each fiscal period, many companies, especially those in the retail trade, will budget a certain amount of money to cover operating losses due to theft of various forms. Such a reserve for theft becomes part of the company's regular operating budget, and the cost

---

[2]Walsh, Timothy J., and Richard J. Healy, *Protection of Assets,* The Merritt Company, Santa Monica, California, 1974.

[3]NAM Reports, *Vol. 21, Number 18, December, 1976.*

of their manufactured product or the selling price of their commodity is adjusted accordingly. Thus, in the end, the American consumer pays the so-called "theft tax" in the form of higher prices. These prices are passed along and added to by each segment of our consumer supply system: the manufacturer, the wholesaler, and the retailer.

The trend has not improved, nor has it even leveled off. Security experts have always agreed that there is a direct correlation between overall crime statistics and the amount of internal theft in industry. In this context, it should be noted that the Uniform Crime Reports issued by the FBI show that larceny-theft in the period from January to June, 1976, was *up* eleven per cent over the corresponding six months of 1975.

Corporate Theft Prevention

In the past ten years, many companies have come to realize that there is more to be gained by taking action other than simply setting financial reserves and adjusting prices to cover the burden of internal crime. With the cost of labor, union settlements, and other aspects of the general inflationary spiral affecting the bottom line or profit figure more and more, decisions have been made to attack the theft problem directly and thus put back into the profit picture some of the funds which had previously been budgeted for write-off.

The past ten years have seen the creation of more and more internal or corporate-level security departments—departments designed not only to keep internal theft in bounds, but to reduce theft through preventive efforts coupled with a direct attack on the perpetrators themselves. Many of these corporate-level departments are now actually viewed as profit centers because of their success in preventing and rooting out entrenched thievery within a company.

Prevention efforts in the setting up of controls to thwart thievery are all well and good and become a necessary part of any well-balanced security program. However, corporate executives should recognize the fact that controls, preventive measures and educational efforts are intended for the sixty or more per cent of American workers who will occasionally yield to temptation, if sufficient management controls are not exercised. At one end of this scale, we can find about twenty per cent of the work force who need no controls and no restraints in order to maintain an acceptable degree of honesty. On the other end of the scale, there is an approximate twenty per cent who generally pay little or no regard to preventive measures or educational efforts and who seem to operate on the philosophy that "whatever I can get away with is mine." People of this type will usually devote their mental energies to determining ways to circumvent controls which have been set up by management, or to taking advantage of certain weaknesses or loopholes which they can spot. For these people, there is only one solution: detection, apprehension, and separation from the company's payroll. The subject of this text, internal loss investigation, is primarily concerned with this dishonest segment of the work force.

Factors in Internal Theft

Undoubtedly, a word of explanation is in order as to some of the reasons for this nationwide problem of internal theft and the expertise which has been developed to counter it.

Population Shifts. Since World War II, great shifts in population within the United States have occurred. A number of factors have contributed to this movement, such as weather and the establishment in specific areas of key industries tied to national defense, the space program, or some other specialized national need. No longer is an employer able to determine the fitness of a job applicant based on his own knowledge of that person and the community from which he comes. More and more, the typical employer has been forced to make his hiring decisions on the apparent ability of the person to perform the job rather than being able to give any in-depth consideration to the individual's character and honesty traits.

Limitations of Reference Checking. Employers have generally become increasingly aware of the severe limitations of background investigation and, in particular, of the traditional reference letter, which means less and less as time goes on. Because of a fear of lawsuits, the overwhelming majority of today's employers are reluctant to give any information in a reference letter beyond dates of employment, job title and salary level. More and more employers have turned, then, to the use of sophisticated screening devices such as pre-employment polygraph testing or a paper-and-pencil honesty test such as the Reid Report.[4] (See Chapter 12.)

Lack of Employee Loyalty. Unlike Japanese firms, which generally enjoy a high degree of loyalty on the part of their employees, the typical American employer has seen employee loyalty dwindle to the point where it can no longer be considered a viable factor in any well-balanced security program.

Much of this erosion of loyalty has been brought about by the labor unions which represent employees in their relationships with their employer. Many labor unions look upon workers as "their people" rather than employees of the particular company for which they work. Many labor locals have also interceded in company internal security matters to the extent that many employer investigations into internal theft problems have been completely disrupted or stopped in their tracks. Union business agents often inject themselves into these affairs by informing employees of "their rights." Many union constitutions or oaths of membership contain provisions prohibiting one union member from informing on another union member. All of these factors have contributed to the almost total lack of employee loyalty found in some com-

---

[4]John E. Reid and Associates, Chicago, Illinois. The Reid Report, developed about 1947, is the original paper-and-pencil test designed to show "attitude towards honesty." It has undergone a number of revisions until its validity and reliability are now accepted by the scientific community.

panies and they complicate further the already complex approach to controlling internal theft.

Restrictions on Screening. Efforts of employers to screen job applicants are also hampered by restrictions imposed by the federal Equal Employment Opportunity Commission and local human rights commissions. After being investigated by either type of body, many employers will drop all pretense of screening job applicants and simply hire people at face value.

Lack of Security Expertise. Also, it must be recognized that the typical American employer today has no particular expertise in the control and investigation of security matters and must depend more and more on professionals in the field.

Organized Crime. Other factors that enter into the complex problem of internal theft are organized crime and its various ramifications. There have been a number of reported instances where organized crime has attempted to infiltrate legitimate business. However, on an overall basis, the problem is not infiltration so much as it is organized crime's capacity to act as a fence for the sale of stolen goods from American business. In any scheme for grand theft of desirable products, ranging over a period of time, the chances are probably better than fifty per cent that some segment of organized crime is involved in the sale of the stolen goods.

Drugs. The growing drug culture in our country today has also had an impact on American industry. We have seen a constant increase in trafficking in illegal drugs and other controlled substances within plants and stores. Many employees regularly conduct sales of all types of drugs to fellow employees on company premises. It is not at all uncommon to discover the use of marijuana and methamphetamine ("speed") on the job.

Gambling. Along with drugs, another theft-influencing factor which has been with American industry for a long time, and which has ties to organized crime, is the problem of in-plant gambling. Many plants and factories today are permeated with some form of the so-called numbers racket.

Larger companies usually have one or more employees on the payroll who will regularly accept and place racing bets on any track in the country. In case after case, the author has found that wherever widespread gambling exists in a company, there is a direct correlation to the degree of internal theft.

Gambling, in this connection, does not include the annual World Series pool or other occasional office lotteries as having an adverse effect on employee honesty. We refer, rather, to the more sinister forms of gambling which can be linked directly to organized crime.

These, then, are some of the more complicating factors that the modern-day employer must face in attempting to control and ferret out theft within his organization.

Difference from Law Enforcement

When considering the foregoing factors which influence internal losses, one might immediately draw the conclusion that the work of the security professional is almost identical to that of law enforcement. It must be pointed out, however, that although the two are similar in nature and share many basics, there are great differences between the law enforcement approach to crime and the security specialist's approach to controlling internal loss.

The first and most prominent difference that may be encountered is that the police have a number one objective of prosecution and, hopefully, removal of the offender from the streets. The security professional, on the other hand, while occasionally resorting to prosecution, has as his number one objective the separation of the offender from the company payroll. For this reason and this reason alone, the police generally do not take an active interest in solving internal theft situations unless there is a concrete commitment by the employer to prosecute.

Additionally, it would be fair to state that most police departments do not have the manpower or the expertise to handle the intricate internal loss investigations demanded of the private security manager. The following case is a good example of manpower limitations.

New York, New York[5] . . . . A high-priced specialty apparel store in New York City suddenly began experiencing obvious losses in men's clothing two and three times per week. A typical discovery was made by sales personnel arriving one morning when they noted that certain items of clothing were missing from the night before. The security department was notified. An investigation revealed a consistent pattern of mysterious disappearances in only the higher-priced men's suits, leather topcoats and jackets, etc. The losses appeared to be in a wide range of popular sizes.

The security department was convinced at the outset that the store was being burglarized at night and promptly made contact with the local police precinct. There was never any sign of forced entry, which made the problem even more perplexing and probably dampened the enthusiasm of the police.

The lieutenant in charge of the local precinct detectives advised that he would be happy to assign two men to assist in an all-night stakeout of the store, but because of manpower problems such assistance could only be rendered for one night. Unfortunately, nothing occurred on the night in question. Therefore, the security department was faced with the need to maintain the nightly stakeout within the store with its own personnel. The police did not lose total interest in the case; they advised that, should the company security personnel succeed in apprehending anyone, they should call the police immediately and assistance would be rendered.

On the third night of the stakeout, the two burglars who had systematically looted the store suddenly appeared on the first floor, coming up through a tunnel broken into the bottom of

---

[5]This case occurred prior to the drastic manpower cuts of 1976 by the New York City Police Department which were dictated by heavy budgetary cutbacks in all city departments.

one of the elevator shafts. The apprehensions were made and the offenders were prosecuted for burglary.

The assistance given by the police department amounted simply to transporting the prisoners to the precinct and handling the paperwork!

SECTION I

INDICATIONS OF THEFT

Chapter 1

Inventory Shortages

Security professionals generally agree that indications of theft almost always come about in one or more of the following three ways:

- Inventory shortages (or variances)
- Evidence or other symptoms discovered during facility inspections.
- Information received.

These indications are the beginnings of a typical internal theft case, and we shall examine each of them in detail in the next three chapters.

INTRODUCTION TO INVENTORY

Many business executives hold to a time-honored belief that inventory shortages are valid barometers of just how a company is doing in relation to internal theft. Many seasoned security experts, however, are quick to point out that, while inventory shortages (occasionally referred to as variances) may be a primary and basic indicator of internal security problems, such shortages (along with overages) are subject to a wide range of variables. Any of these can influence the picture to such an extent that normal conclusions may often be completely invalid.

It is not the intention of this text to cover in anything but a superficial way the accounting complexities which are involved in inventory problems. For those approaching the subject with no training in accounting, however, some primary explanations of this very important aspect of loss prevention are essential.

Book and Physical Inventory

Basically, there are two kinds of inventory: the "book" inventory and the "physical" inventory. The book inventory is arrived at as follows: At the beginning of each fiscal year, a company will have arrived at a figure of its inventory on hand by adjusting its book inventory to agree with the value of the physical inventory counted at year-end. Throughout the year, certain purchases or production of additional inventory are made by the company. When received, these amounts are added to the beginning inventory figure. Also, sales of the company's products are made throughout the year, reducing the level of the company's inventory. Therefore, these sales must be subtracted from the other figures which are normally augmented by purchases or production.

At the closing of the fiscal year, the company has thus arrived at a certain figure which represents the beginning inventory, plus purchases and production, minus sales. This figure is known as the book inventory.

Once the book inventory is determined, virtually all companies are required to then make a physical count of all the inventory in their possession or under their control. This physical count may take several days and involve the efforts of many people. The figure resulting from the count gives the "physical" inventory. This physical inventory represents not what the company thinks it has on hand, but what it actually does have on hand. This count is compared to the book inventory.

Invariably, there will be unit differences between the physical inventory count and the book inventory figure. These differences can take the form of an overage or a shortage. Nominal differences or variances usually receive absolutely no attention, as they are considered normal for any business. Significant variances, however, whether they be overages or shortages, should be of concern to management.

Variances which are reflected as overages can be traced to a number of factors, such as price increases in commodities which went unrecorded in the books during the fiscal year, and also to poor physical counts.

Shortages can also result from a variety of causes, including internal theft, undiscovered external thefts, human error, and, once again, incorrect physical counts. Other variances can be caused by incorrect compilation of the book inventory amount and improper cut-off procedures.

Possible Weaknesses in Physical Inventory

It would seem clear, then, that any security executive attempting to study and analyze the results of a particular inventory should always be concerned with the adequacy of the actual mechanics of the physical count. There are a number of good accounting texts, such as *Retail Accounting and Financial Control,*[6] which cover the proper methods of taking a physical inventory.

The security executive should also concern himself with possible weaknesses in the system which have been caused by deliberate actions on the part of one or more members of the physical inventory team. One question that should be asked is whether or not the primary counters of the inventory are the persons who could have been responsible for theft of that particular portion of the inventory which they are now attempting to count. In numerous cases, key personnel have attempted to cover thefts throughout the year by padding inventory figures.

---

[6]Moscarello, Grau and Chapman, *Retail Accounting and Financial Control,* Fourth Edition, Ronald Press, New York, New York, 1976.

In most inventories, double counts are taken. It is essential that the security executive or accounting official make sure that there is no form of communication between the primary counting team and the second counting team. There have been occasions where the primary team, when finishing the count of a particular section of finished goods, would leave a number signifying the first count in a predetermined location in the section. This was done as a "favor" to the two men on the second team so that they would not have to make an actual physical count themselves.

As most inventory teams are constituted, each member of the team makes initial counts in certain sections and then serves as a secondary counter for other sections. If all of the primary counters leave some kind of communication behind as to their count, then it becomes evident that no actual second count is ever taken. In this way, discrepancies between the first count and the second are never noted; thus a third and final count by members of management to verify a particular discrepancy is never made.

It is not difficult to visualize how an attitude of indifference on the part of certain workers could give rise to major inventory errors. One of the greatest deterrents to lax counting procedures by local personnel is a test count made by internal or external auditors. The mere presence of an auditor during the taking of a physical inventory certainly must have a beneficial effect. An auditor who takes nothing for granted and is aggressive and inquisitive by nature adds significantly to the integrity of the whole process.

INVENTORY IN MANUFACTURING

In manufacturing, there are two primary types of inventory with which the security executive must concern himself. These are raw materials, and goods in the production line which are referred to as "in-process." (Finished goods, which may be stored and warehoused at a factory location, will be covered under the heading "Distribution and Warehousing.")

Raw Materials

In the case of raw materials, the basic accounting steps are similar to those outlined earlier. At the beginning of the fiscal year, the company starts out with a certain amount of raw materials. Additional raw materials which are received by the company throughout the year are added to this amount. Those amounts of raw materials which are withdrawn for purposes of production are then subtracted from these figures.

If there is good record-keeping and a sense of responsibility on the part of workers and supervision, the difference between book inventory and the physical inventory balances should not be large. At least it should be within acceptable tolerances, as there will always be a certain amount of waste in the production of goods from raw materials.

Accounting procedures have developed a standard as to how much waste is tolerable, and therefore this is not too difficult to measure. "Standards" are used to record consumption, but any usage in excess of standard is recorded as a cost variance (not an inventory variance).

The following case, which involved two widely separated manufacturing locations, is an example of how an adverse human element may affect a manufacturer's inventory figures.

Saratoga, New York . . . . At a cloth mill which produced and processed cloth for overseas manufacturing, the night shift workers (and supervisors) seemed to be more intent on card-playing, drinking on the job, absenteeism, and general horseplay than they were on the task of faithfully producing the company's product.

Undercover investigation revealed that on numerous occasions, after a drinking bout, one or more of the workers would kick a defective totalizer on a certain cloth processing machine. The totalizer recorded the number of yards of cloth produced or run through that machine, and this figure was used to pay the workers for piece rates. It was also used in preparing shipping invoices to the overseas customer, to determine the total number of yards on a particular roll of cloth which was being shipped.

The workers discovered that kicking the defective totalizer would usually make it "jump" by 1,000 yards. In this way they would be able to show relatively good production for an eight-hour shift during which they might have worked only five or six hours.

On the receiving end of this operation was the overseas facility which received the yards of cloth. Not attending to details properly, the receivers would simply accept the yardage quoted by the cloth mill as being bona fide. A false figure was thus generated as to how much raw material had been received and was in inventory.

In-Process Goods

An in-process inventory is exactly what it implies: that amount of unfinished goods which are in the production pipeline at the time the plant is shut down for the annual physical inventory. This amount of goods is not normally excessive; it should be only, as indicated above, that amount in the manufacturing process on a particular day. Finished products coming off the end of the assembly line are often not recorded until a count is made for shipment into a central warehouse or distribution center.

Although in-process inventories have limited impact on financial statements and security considerations, they are also highly dependent on accurate production records and the integrity of the workers on the production line. Where component parts are being pilfered by production workers, many companies have instituted programs of lot control. Lot control, in its simplest form, means that at each major stage of the production system a particular lot must be fully accounted for before being passed on to the next major section of the production line. This means, theoretically, that there would be no missing components. In the garment industry, lot control has become of major importance in controlling in-process inventory shortages.

DISTRIBUTION AND WAREHOUSING

In distribution or warehousing, there are two major considerations as to inventory. First is the finished goods inventory which results from the same basic accounting procedures discussed previously. Finished goods in a typical warehouse or distribution center should be influenced only by transactions that take place in the receiving and shipping departments. In other words, nothing should be added to the book inventory until it is actually physically received by the distribution center, and nothing should be subtracted from the book inventory until it is actually physically shipped out of the distribution center.

Concealed Shortages

The second area of concern to the security executive, and one which is not normally accounted for by standard inventory figures, is concealed shortages. A concealed shortage can be defined as a shortage of goods or inventory which is not obvious to the examining party.

Take as an example a liquor warehouse where cases of Scotch are being received. If a case has been tampered with and one bottle of Scotch removed, and the case shows no evidence of tampering, a concealed shortage exists. This may be discovered during the course of the fiscal year if the liquor warehouse has occasion to break open the case for the purpose of selling individual bottles to its customers. If this happens, the shortage is then noted and an appropriate adjustment can be made on the book inventory figures so as not to cause a physical shortage at the end of the year.

On the other hand, if the shortage is not discovered and the case is sold and shipped intact to a retail customer, the shortage is then discovered by the retailer. At this point the retailer asks for a credit adjustment on his account with the wholesaler. The credit adjustment is usually made, but the adjustment does not have any effect on the wholesaler's physical inventory. Unless a separate record is kept of all credit adjustments made throughout the year for concealed shortages, the security executive never obtains a true picture of what is transpiring in a particular distribution center.

In the apparel industry, the same type of situation can exist where items are normally packaged three or four to a box. When the order-picker is filling a customer's order and withdraws one item from a particular box, it is general practice for him to adjust the number on the outside of the box so that a subsequent order-picker will know that the box does not contain the full allotment of garments. If this is not done, through negligence, then the customer is shorted and credit adjustment must be made. In the case of pilferage from these boxes, the thieves usually do not bother to make notations of the quantity reductions in the box, and thus concealed shortages are created.

Of course, security executives will immediately recognize that the problem of concealed shortages may very well be on the other end—in the retailer's receiving room. For this reason, spot checks must frequently be made of boxes in a warehouse as well as occasional double verified shipments to customers. Generally speaking, however, the accounting and security departments must be guided by what they feel are "standards" for their industry in the matter of concealed shortages. Oftentimes, deviations from the standard can be the tipoff that something is seriously wrong, as in the following case.

A moderate-sized shirt company was experiencing what appeared to be excessive concealed shortages in shipments of shirts to customers. As a standard, this particular company was compared to a larger shirt company within the same corporation that manufactured approximately three times the volume of the company in question. When these comparisons were made, it was found that the moderate-sized company had twice the amount of concealed shortages as its big brother, which was three times its size and in an identical business. With this kind of comparison, one can only conclude that there are major problems within the distribution or warehouse system itself.

RETAILING INVENTORIES

The very nature of the retail business dictates that inventory control and analysis are far more complex than with either manufacturing or distribution. In retailing, throughout the year there are annual sales, special sales, and many other factors which, if not recorded and handled properly, will have an adverse effect on inventory figures.

Shrinkage and Reserves

In retailing, inventory variances or shortages are referred to as "shrinkage." In virtually all retailing (and in some manufacturing companies as well) there is a psychological attitude that a certain amount of shrinkage is inevitable. As a result, the companies actually budget for a certain percentage of shrinkage throughout the year. In so doing, if their forecasts are correct, they do not close their books at the end of the year with a loss in inventory—a loss which would, of course, affect the bottom line of any profit-and-loss statement.

These shrinkage reserves vary from store to store and from one segment of the retail industry to another, depending on the type of business. For instance, in the conventional department store, reserves may be set up which provide for inventory losses of from one-and-a-half to three per cent of the net sales. In the discount or mass merchandising segment of the retail industry, these reserves usually range from two to four per cent. In clothing or specialty apparel stores, such reserves will usually range from one to two per cent.

Shortage Reduction As Contribution to Profits

Some security directors have targeted their programs towards these reserves in an effort to bring the inventory results in under the figure which has been reserved. If the actual shortage is lower than the reserved shortage, the difference can rightly be claimed as a direct contribution to profits of the company. In other words, the company is bringing back into the profit column a certain portion of those reserves which were not utilized in covering for shortages which took place during the year. One major mail order house which also operates a large number of retail outlets has always taken great pride in announcing this actual favorable difference as a contribution to the profit picture of the company.

Theoretically, if buyers, merchandise managers, and other executives did the proper, error-free paperwork; if sales clerks did not make mistakes with customers; and if all internal and external thefts were completely obliterated, then there would be absolutely no so-called "shrinkage." Accordingly, in the retail stores division of the author's company, Cluett, Peabody & Co., Inc., it has been publicly stated on many occasions that the goal of the specialty apparel stores in that division is a shrinkage of less than one per cent. Several of the store groups within that division have achieved that goal for two or three consecutive years.

Shortages Based on Daily Count

In some segments of the retail industry, stores sometimes make a daily count of major clothing items. These counts are normally made at the end of each selling day and should correspond to the amount of sales of a major clothing item such as women's dresses or men's suits for that particular day. Where problems are known to exist, an additional daily count is occasionally performed at the beginning of the selling day as well.

From time to time, these daily counts, although accurately reflecting the amount of shoplifting at a particular location, also may indicate problems within the store after closing. The following case history is an example of how a daily count inventory may add to the woes of the security manager.

Sherman Oaks, California . . . . Over a period of several months, the daily suit count of a clothing store showed significant discrepancies on an average of once or twice each week.

In attempting to pinpoint the discrepancies, the security manager instituted an additional count at the beginning of each business day. The additional count continued to confirm the previous pattern and indicated that suits were disappearing from the store during the closing hours.

The security manager undertook a number of lines of investigation, including a secret audit of the alarm company records, surveillance of the store, installation of a hidden closed circuit television camera which was designed to operate during the nighttime hours, and other measures. None of these efforts, over a period of weeks, revealed anything, nor did a minute examination of the actual alarm system and physical construction of the building by an expert brought in for that sole purpose.

Next came a further review and audit of the daily stock counts. It suddenly was apparent that discrepancies were occurring in the morning counts that were being taken by a relief manager who only worked in the store one or two days each week. The morning count would be at variance with the preceding evening count which had been taken by the regular store manager or his assistant. On those days, the evening count would always balance with that day's morning count.

Interrogation, coupled with polygraph examination, revealed a pattern of dishonest salesmen and the assistant manager padding certain customer purchases. Padding was accomplished by adding extra garments to legitimate items being purchased. The resultant shortages were then covered up by the store manager in the daily counts which he made, during which he "forced" the balances.

Although he was not personally involved in the thefts, the manager could not bring himself to admit, through his daily count, that there was anything wrong in his store. The discrepancies came to light only because the relief manager was making a completely honest count. In the meantime, thousands of dollars of men's suits had been lost through collusion. In addition, another considerable amount of money was expended for unproductive investigative efforts.

INVENTORIES BASED ON ESTIMATED GROSS PROFIT

In many companies in the United States today, book inventories are not maintained in the manner described in this chapter. In the distribution field, because of the large amounts of goods which are warehoused and distributed, upward and downward price changes are not recorded on the books throughout the year.

One major distributor with nationwide operations, for example, stocked from twenty to twenty-five thousand different items of merchandise. It was virtually impossible to accurately record price fluctuations in book inventory throughout the year on every one of these products. Therefore, the company operated on what was termed an estimated gross profit. Estimated gross profits are simply arrived at by evaluating the actual gross profit percentages realized in prior years, and are based on what an intelligent appraisal of the company's business indicates will be the amount of gross profit.

"Cost of Sales" Category

In the manufacturing field, a similar estimate is sometimes referred to as a "standard" by which the company arrives at its gross margin of profit. In these situations, the basic accounting formulas described earlier are followed, except that a term known as "cost of sales" comes into play. Cost of sales in a manufacturing operation can be defined as all manufacturing costs incurred in finishing or producing a product. This would include raw materials, direct labor, and overhead costs. In wholesaling or distribution, cost of sales would usually include purchase price of the merchandise and any related transportation costs.

Where there are no proper records being kept, price fluctuations or shortages in goods from the manufacturing unit to the warehouse are simply absorbed in this accounting category known as cost of sales. When this system is utilized, the only standard by which a company can judge its performance is an estimated gross margin or gross profit, based on past performances.

Built-in Theft Factor

The security executive must be acutely aware of the fact that, where estimated gross margins are utilized, there is always the possibility of a built-in historical theft factor. As an example, when the author first joined McKesson and Robbins, Inc., the company's drug department was operating, and had always operated, on an estimated gross profit margin of about seventeen-and-a-half per cent. After a security program was instituted and some long-term entrenched thieves were rooted out, gross profit margins started to rise suddenly to the area of eighteen and eighteen-and-a-half per cent. This was astounding, considering the fact that for many years officials of the company had felt that a gross profit margin of seventeen-and-a-half per cent was a normal standard for their business.

In other words, if there has been a level of theft which has remained fairly constant throughout the years, there would be no inventory shortages. It is only when people become greedy and begin to steal more than the historical amount that inventory shortages are discovered. The following case history is an example of an unusual situation where even an accelerated theft program did not show itself.

Seattle, Washington . . . . A theft ring of six men in a drug warehouse was suddenly discovered in January. The ring had been in operation for only six months and had been engaged in selling stolen pharmaceuticals to several drugstores in the metropolitan area. The ring had, by their own and two druggists' admissions, succeeded in smuggling out approximately $10,000 in pharmaceuticals in the six-month period.

The company closed its books for the fiscal year on March 31 and at that time took a physical inventory. Because of the fact that the company was operating on an estimated gross profit, not only did the $10,000 in pharmaceuticals not show up as an inventory shortage, but the company actually showed a small overage on its books!

VARIANCES NOT CAUSED BY THEFT

As we have indicated indirectly throughout this chapter, there are many factors that can adversely affect inventory records, causing either overages or shortages. A comprehensive review of inventory problems must include some of the more common causes of variances that do not involve theft.

Price Changes

In the retail field, the changing of prices throughout the year is known as taking "mark-downs." These must be recorded in the books. Failure to record mark-downs properly in the book inventory will invariably cause shortages at the end of the year. Other price change considerations include mark-down cancellations, mark-ups, and mark-up cancellations. All of these procedures, if not properly recorded in the inventory books, will affect the book inventory at the end of the year in an up or down manner.

Receiving Short Shipments

Another vulnerable area in any business operation is the receiving room. Security executives are well aware of the practice of a dishonest truck driver attempting to divert the attention of a receiving clerk so as to retain one or more cases on the truck while the receiving clerk is led to believe that the count is accurate and that he has received the correct number of cases in the shipment. This practice is sometimes referred to as receiving short shipments.

Also, there are frequent occasions when a receiving clerk will discover that a shipment is short but will not make note of it, failing to notify the accounting department accordingly. When this happens, the full shipment is put on the inventory books and an instant shortage is created.

A variation on this theme sometimes occurs whereby goods

are received and then subsequently counted in a physical inventory but are not recorded on the books until after the physical inventory is taken. This, of course, creates a "cushion" and can help to cover up other wrongdoing which has taken place during the year. By the same token, however, it also creates an instant shortage in the subsequent year.

Breakage and Damage

Another area not to be overlooked is breakage and damaging of goods by rank-and-file employees. Many employees, wanting to avoid reprimands by their supervisor, will fail to report broken and damaged items and will simply discard the evidence into the trash. Where breakage and damage are not properly recorded so that proper book adjustments can be made, once again the inventory is shorted.

Unintentional Error

Last, but certainly not least, is human error. Indeed, it is probably one of the biggest factors of all. Unintentional underringing of sales to customers, the miscounting of goods in a customer's shipment, errors in typing, and endless other examples could be given of human error which can cause inventory shortages.[7]

ADVANTAGES OF DATA PROCESSING SYSTEMS

In many companies today, the security executive can avail himself of a data processing system to assist in determining trends in pilferage, etc. Many of these computer systems are used by companies for buying purposes only, and although they are not used for actual accounting records, in many cases they do constitute a perpetual inventory system. That is, on any given day, the computer may be able to indicate to a

---

[7]Additional specific examples of this problem are dramatized in an excellent film on the subject, entitled "Who, Me?" produced and distributed by Anne Saum and Associates, New York, New York.

buyer or the security executive how much of a particular item of stock should be in the possession of the company and available for sale.

By simply monitoring the actual sales records of a particular item and then comparing them with the print-out of the data processing system or perpetual inventory, the security manager is often able to verify suspicions and confirm that theft is currently taking place. It goes without saying that data processing systems offer tremendous advantages in the control of internal theft as opposed to waiting for results of the annual year-end inventory.

Chapter 2

Facility Inspections

Most sophisticated security programs today include an on-going plan of security inspection of the company's various facilities. Admittedly, most of these inspections are devoted to a review of procedures in order to determine compliance with company policy. The aggressive security manager can, however, utilize these periodic security inspections to look for indications of theft within the operation.

It would be impossible in the confines of this text to outline specific indicators for every type of business. Each type of industry is different, and only by learning in detail the nature of a company's operation can the security manager hope to be able to arrive at certain indicators of theft. For the purpose of illustration, however, we shall attempt to point out common basic points of inspection plus some specific examples of vulnerable areas in certain specialized industries. From these examples it should be possible to develop some insight into how particular guidelines or indicators of theft can be compiled for a particular company.

INSPECTIONS IN MANUFACTURING AND DISTRIBUTION

Records of Openings and Closings

Many facilities today are covered after-hours by an alarm system which may be monitored from a central station facility. Where central stations are utilized, it is almost always possible to arrange to receive reports from the alarm company of any unusual openings and closings outside of normal hours. These are often supplemental reports prepared independently of the regular daily opening and closing reports provided by the alarm company.

In the case of installations with proprietary alarm systems, these systems may come equipped with an automatic printer or recorder which indicates not only the openings and closings of any doors in the installation, but also any tampering with the console or the printer.

For locations that cannot be covered by a central station system or a proprietary system with a built-in recorder, a time-recorder lock can be installed on the main entrance of the facility. The time-recorder lock in turn is tied to sequence locks on other doors. In this way, the security manager can obtain the same permanent record of unusual openings and closings offered by more sophisticated systems. This record can be invaluable when he is looking for a particular pattern of surreptitious entries or exits at the facility.

In addition to studying records of openings and closings for obvious holiday, Sunday, and nighttime openings, a security manager should also be alert to the extra two- or three-minute opening that occurs just after closing of the facility at the regularly scheduled time. Often, dishonest personnel have been known to place a cache of contraband near the exit and then re-enter the building within several minutes of its closing on the pretext that they had forgotten a hat, umbrella, handbag, or some other item. This pretext is usually employed for the benefit of a fellow worker who may have left the facility at the same time as the key-holder.

Where guards are on duty and change shifts at night, the situation becomes far more complicated. It is quite a common practice for some guards to arrive at their work place anywhere from thirty to sixty minutes in advance of the relief time, and this may or may not precipitate an early departure by the guard being relieved. The problem, of course, comes about if there is collusion between the two guards and contraband is removed during the authorized opening of the door for the changing of the shifts. One solution for this possibility is periodic surveillance of the facility at these times.

Employee Locker Inspections

Employee lockers are another very productive area which should be inspected in detail. Although it is unusual to find any hard evidence of theft in such locations, the security manager can frequently pick up other clear-cut indications of violations of company rules and regulations. Often, the major violators of certain rules can also be identified later among the dishonest employees of a location.

Alcohol. In particular, inspectors of lockers should be alert to alcoholic beverages, not necessarily in the original container. Many times, alcoholic beverages have been found in ordinary glass bottles, plastic containers, and in thermos jugs.

Drugs. By the same token, many of the commonly used commercial drug preparations can often be found secreted in employee lockers. Here, though, the security representative must keep in mind that it is always possible the employee has a legitimate prescription for the drugs. This would be especially true for the so-called "uppers" (Dexedrine, etc.) or "downers" (barbiturates). On unidentified substances, the security manager should make an attempt to remove enough of the substance for laboratory analysis but at the same time should attempt to photograph the remainder. He should also note the time, date, place, and any witnesses to the search. The most usual unidentified substance to be encountered will be the so-called "nickel" or "dime" bags of marijuana, or possibly cocaine.

Gambling Items. Gambling records or paraphernalia may also be found in lockers. There may be memorandums of I.O.U.'s or rosters of players within a gambling pool where identification numbers are being utilized. On occasion, the paraphernalia for various athletic or payroll check pools may be found.

Evidence of Theft. As stated earlier, it is the exception rather than the rule that hard evidence of theft will be found in lockers. But sometimes packaging materials may be discovered which would indicate that a finished product had been appropriated for personal use. The most common disposition of such materials is usually into a trash container. On occasion, certain wrappings or other identifying labels are discovered in employee lockers, especially if there is some indication that trash containers are being checked by management on a daily basis.

Legal Considerations. Under the existing body of United States labor law, the rule has been laid down clearly in case after case that, where an employer puts his employees on notice that lockers are subject to inspection and will be inspected, it becomes a management prerogative to do so. There is absolutely no question of invasion of privacy or unlawful search involved at this point.

The one consideration that may play a part in determining the method used for locker inspections is whether or not certain agreements have been made with the plant union on this subject. Some agreements provide that the employee must be present when his locker is inspected, and if he is not, then the fruits of any inspection cannot be held against the interests of the employee. In large plants, where it would be impractical for every employee to be present during a locker inspection, a compromise can often be worked out with the union to provide for a shop steward to act as an employee representative during an inspection.

At any rate, regardless of representation by or on behalf of the employees, locker inspections should be conducted without any prior warning and at irregular intervals. Private padlocks should be prohibited; lockers should be provided with locks that have duplicate keys for management or even a master key.

Many security managers, not wanting employees or unions to become alert to the discovery of theft, will conduct their own private locker inspections after working hours, in a surreptitious manner. The object here, of course, would not be to make a case against a particular employee, but only to give the security manager some intelligence as to what may or may not be happening in the plant. Decisions on official inspections, which by their very nature precipitate action against offending employees, should be made in a careful and considered way.

Work Stations

In most factories and distribution centers, employees have a particular area where they tend to accumulate minor personal effects over a period of time. Generally, this could be referred to as the employee's work station. At many work stations, there are desk drawers to be examined, the rear of file cabinet drawers, small boxes containing personal items, or other containers. Here again, the security manager should be alert to any evidence of gambling, packaging material, contraband merchandise, or any form of illicit production records such as false shipping labels, false work orders, etc.

Sensitive Areas

Many plants and distribution centers have sensitive areas to which particular attention must be given during a security inspection. This would include special security rooms, security cages, and similar areas. Not only should desk drawers or file cabinets in areas of this type be closely examined, but also waste containers, as the following case history illustrates.

Cleveland, Ohio . . . . On a routine annual security inspection, a security manager inspected a security cage and noticed approximately six cardboard sleeves which normally were used to cover wristwatch boxes. Closer examination of the trash also revealed a number of empty watch boxes.

The discovery was made without alerting the order clerk. Therefore, it was possible to brief an undercover agent who had been assigned to the distribution center. The agent was able to gain access to the security cage, and he checked the trash container for several days, both at lunch time and at quitting time. It quickly became evident that an illicit traffic of from twelve to twenty-four stolen watches was in progress every day.

From here, it was a simple matter for the security manager to apprehend the order-filler who worked in the security cage as he departed the distribution center for lunch. At the time of his apprehension, he was wearing approximately twelve new watches on his arms. He was arrested and charged with grand larceny.

Lavatories

Inspections of lavatories, both men's and women's, are not to be overlooked, as they often prove productive in the revelation of indicators of theft. Particular attention should be given to toilet water tanks, not only the inside of the tank but also the space between the rear of the tank and the wall; the area underneath wash basins; and any small access doors to plumbing traps, etc. Shelving, ventilation duct work, dispensers of various types, as well as all types of refuse containers should be examined closely.

In a facility which stores or warehouses items of clothing such as hosiery, panty-hose and underwear, the inspection should be particularly alert to old items of this type of clothing which may have been discarded in trash containers. It is quite common for women to discard worn panty-hose in trash containers in the women's lavatory and simply replace them with a new pair from stock. Usually, along with the old discarded pair, the security manager can also find the original packaging material in which the new product was contained.

Full Case Area

In distribution centers or warehouse areas, any security inspection should pay particular attention to the full case area. If full cases are utilized for picking purposes in the filling of customers' orders, normally such cases will have been removed from the full case area and can be found in an open stock or order-filling area. The security manager, then, can conduct a meaningful inspection of the remaining full case area.

Cases in the full case area should be intact and sealed. Dishonest employees will sometimes deliberately damage a full case with the forklift truck or some other object merely to make it easy to retrieve contraband from the container. The security manager should be particularly alert to any apparent damage to the cases to insure that the interior contents are still intact. In other areas of distribution, where customers' orders may be processed on the basis of three or four items to a smaller packing box, checks should be made of these containers to insure that no concealed shortage exists.

Miscellaneous Areas

Other areas that should not be overlooked by the security manager during his periodic inspections are crawl spaces between buildings, air shafts, elevator pits, stairwells, roofs, utility or steam tunnels, etc.

In general, the security inspector should be watchful for any indications whatsoever within the facility of the existence of gambling, drinking on the job, or use of drugs. He should be alert to employees who appear to leave their work places for unauthorized reasons or proceed into unauthorized areas. Also, sleeping on the job as well as horseplay are often good barometers of the moral atmosphere within a plant. Many ingenious employees have devised methods of stacking full cases of merchandise so as to provide a hollow area within the stack in which they can drink, play cards or just plain sleep during the day. The security manager who knows his installation knows the practices of such people, and he can usually succeed in discovering their secret hiding places.

INSPECTIONS IN RETAILING

The retail store has its own special peculiarities to consider during a security inspection. As in manufacturing and distribution locations, these retail peculiarities can give clear-cut indications of possible theft or rule violations which give rise to eventual theft. The specific type of retail establishment will determine the particular indications of problems with which the security manager must be concerned. There are common denominators, however, and it is useful to examine some of the more generalized aspects of what to look for in such an inspection.

As in manufacturing and distribution, areas of concern in retail security inspections include merchandise which is hidden in unlikely places, poor housekeeping, and of course a pattern of unusual opening and closing of the facilities, based on alarm company records.

Another area not to be overlooked, and one which would apply to any type of business, is any indication of tampering with the alarm system. Detection of tampering requires actual testing of every opening in the building with the alarm company.

Surprise Cash Counts

Most retail security experts agree that the first item of business during a security inspection should be a surprise cash count of the various funds which are maintained in retail stores. Here, the security manager will be on the lookout for any shortages, in addition to post-dated checks, I.O.U.'s and other unusual items found occasionally in cash drawers.

The security review should also include a study of the house accounts of the employees to determine any significant delinquencies. An attempt should also be made to ascertain whether certain employees have had wage garnishments or have made repeated requests for salary advances. All of these items can indicate that certain employees are in financial difficulties and should bear closer scrutiny on the part of the security manager.

"Hold" Merchandise

Virtually all retail companies have a stated policy on customer or employee "holds." Simply stated, this is merchandise which is desired for purchase by a particular customer or employee at a later date. It is set aside, usually in a rear stock room, and held for a stated period of time. Upon the expiration of this period, the merchandise should be promptly returned to stock for sale to someone else. In order to control this type of situation adequately, most retail companies have specified that "holds" are to be clearly identified with the customer or employee's name and the date on which the "hold" was taken from stock and placed in a non-selling area. The security manager then should carefully examine all of the "holds" bearing no names or dates or those with grossly expired dates.

Cash Register Irregularities

The retail security manager should be especially alert, during his security inspection, to "no sales" showing on a cash register window—or, for that matter, any opening of cash drawers for no apparent purpose. This would also include the discovery of a cash drawer in an open position with no employee in immediate attendance. The security manager should also be alert to any blatant failure on the part of certain employees to follow house rules.

Proper Documentation

Two final areas which must be mentioned are a review of outgoing merchandise to determine whether any of it lacks proper documentation; and examination of merchandise in alteration departments to see that it has proper documentation.

Summary

In summary, the enterprising security manager in any type of business can utilize his security inspections not only to improve and maintain control procedures, but also to determine clear indicators of theft. In many cases, he may be able to identify specific employees who are involved in some form of dishonest program.

After inventory shortages, then, the second method of detecting theft within an organization is the facility inspection. To these two indicators can be added a third: information received.

Chapter 3

Information Received

From Anonymous Letters and Telephone Calls

Most companies today periodically receive anonymous letters and telephone calls from outsiders. Some of these communications deal with complaints, some are veiled or open threats to the company. Others report crime or other internal wrongdoing. These reports are made by persons who may have good intentions but for reasons of their own choose not to become involved openly.

The task of the security director is to set up channels so that anonymous letters or telephone calls which deal with crime within the company are routed to his attention. In a large national corporation, the letter writer or caller often does not know to whom his communications should be directed. Consequently, such communications are often directed to a local plant and may easily wind up being received by someone in middle management at that location. The security director must undertake an educational task to insure that information received in this fashion finds its way to the corporate security office.

At the outset, it must be recognized that anonymous letters and telephone calls dealing with security matters can have a variety of motivations. Knowing this, the security director must realize that some communications will be close to one hundred per cent truthful, others may contain only partial truths, while others may be completely untruthful. A communication of the latter type can be motivated through personal hatred of one person for another. In some rare cases, certain false communications are designed simply to lead the security department on a wild-goose chase.

Each letter or telephone call must be analyzed carefully. The prudent security director of a national corporation will usually attempt to get some input into his evaluation attempt from a trustworthy key executive at the location in question. In spite of this, it is still possible to be misled by a communication if it is cleverly put together and takes advantage of local information and practices. The following case, involving both a letter and a phone call, will give a better idea of the type of communications which can be received, along with action taken.

Upstate New York . . . . In March, 1975, a hand-written letter addressed to the president of a major division of Cluett, Peabody and Co., Inc., was received at the division's executive offices in New York. (The hand-written letter is reproduced in printed form.) Exact names and locations have been omitted in order to save embarrassment to those persons implicated.

> Cluett and Peabody
>
> I'm might fed up to my ears with a bunch of rowdy jerks who come into my grill time after time trying to sell my customers shirts stolen from your Cluett's mill in ______. I refuse to be an accessory to their shady deals with stolen goods and I won't have my customers annoyed by these damn fools any longer. My wife and my kids say it's best to tell officials where the shirts are stolen from, maybe Cluett can put a scare to employees. It might stop and maybe my customers won't be pressured

to buy stolen Cluett shirts any more. If I turn them in to the cops for sure these rowdies will burn my grill to the ground. Besides the cops bought some of their stolen shirts. They have the cops in their hands. M____, who claims he is head security guard at Cluett, ______, and old man G____ and P____ are tough eggs. They have bad reputations around grills. When the old rowdies get together in my grill, all hell lets loose. They have some gall using my place to drum up customers to sell their stolen loot of shirts to. They interfere with my business and annoy my customers.

They've got some kind of shady deal going for themselves with some jane they call P____. She helps them sell shirts, according to P____. He said this jane P____ works with his wife in Cluett's and told me if it wasn't for his wife and this P____ woman stealing a bunch of piece price tickets from a floorlady's desk safe, that old man G____'s wife wouldn't have all the money G____ says his wife has. G____ likes to brag. P____'s wife and this P____'s woman must have stolen big loot, because G____ says his wife is doomed to paying it back to P____ with big favors. M____, your Security Guard, and these other two birds P____ and G____, sound like they have some big loot of stolen shirts. They claim that over the years their wives have stolen a shirt every day. They brag that they'll be living on the profit they expect to get off the sale of shirts they have stashed away on reserve for the rest of their lives. I don't want these rowdy, aggravating bitches selling their stolen loot in my grill. Customers back out the door when they see these bitches pressuring for a sale.

My wife and I are nervous wrecks. These Cluett crooks are hurting our business. We had it with you so-called Security Guard M____ and G____ and P____. Like I said, if Cluett put a scare to their employees it might stop your crooks stealing shirts and maybe my regular customers can come to my grill again for relaxation in-

> stead of Cluett's shirt sale pressure. The P____ and G____ women work in Cluett's laundry, ______. M____ told my wife.
>
> Discouraged Grill Owner

After conference with key local officials, it was decided that the letter seemed to be logical in nature and correct as to identification of persons and types of jobs which they held with the company. A quick preliminary investigation confirmed the possibility of the theft of piece rate tickets as stated in the third sentence of the second paragraph of the letter.

Ordinarily, this preliminary confirmation, along with other verifications received from key executives at the location, would have been sufficient to trigger investigation by the security department. In fact, preliminary plans for just such an investigation were taking shape when the location referred to in the letter received the following anonymous telephone call. The call was received by a middle management executive approximately sixty days after receipt of the letter. The executive wrote a quick hand-written memo following the phone call and passed it to his superiors. It was then relayed to the corporate security office. The gist of the phone call, reproduced from the hand-written memo, was as follows:

> Following is a summary of a telephone call I received at approximately 11:20 a.m. on Thursday, May 22, 1975.
>
> The caller, a woman, asked our switchboard operator for Mr. K____. The operator directed the call to me, a common practice since Mr. K____'s transfer to Pennsylvania. The caller said she wanted to "report a grand larceny," and went on to tell me about a man we have working for us in ______ who steals shirts on a regular basis and who has been doing so for some time, making a regular business of it. She said he wraps the shirts around his body and leaves by the garage door; the shirts are picked up by a man from New York City named Mr.

G____ (she could not spell it; this is the way it sounded to me), who pays our employee by check.

The caller went on to say that this employee should not be working when so many honest family men are out of work. She said this man has, over the years, molested many of our young female employees and has impregnated a few.

Although the woman was very well-spoken and had, in my opinion, thought out what she was saying, there was a certain emotion in her voice. I asked her if she had any personal reason for calling. She said "yes," that her daughter was pregnant by this man. I asked if she or her daughter planned any legal action against this man. She said "no." The caller identified the employee as R____ S____.

Early in the call I asked the woman for her name and again after she spoke of her daughter. She declined, saying that we could catch him if we watched him.

A. ________
May 23, 1975

A massive investigation was begun shortly after the receipt of the telephone call. In the initial stages, investigators took the position that they were dealing with two separate cases at the same location.

Many months went by during an effort to corroborate the allegations raised in the initial anonymous letter, including an attempt to identify the bar and grill in question. Although there were twenty or thirty likely prospects in the area, it was never possible to identify a bar and grill which was operated by a husband-and-wife team and which was frequented by any of the suspects identified in the letter.

Concurrent with the investigators' attempts to identify the bar and grill, tail jobs were conducted on various persons mentioned in the letter and surveillances of several of the homes were carried out. Over these months, never once did any of these persons have occasion to get together or to meet. In fact, all of the people mentioned in the letter lived in widely separated parts of the community. They all had their own diverse social activities, and none of them ever came together in the evenings or after work. Some of the personalities involved were probably not even compatible, according to the results of the various surveillances.

On the other hand, the investigation of the suspect identified in the anonymous phone call was as productive as the investigation of the various suspects identified in the letter was unproductive. In fact, surveillance of the subject of the phone call led to the identification of other thieves in the plant. With the help of an undercover investigator, and extensive surveillance by additional investigators of the plant, automobiles, and suspects' homes, a very successful case was put together, one which identified approximately fifteen persons as being involved in heavy systematic thefts from the company.

The undercover investigation tended to clear the original subjects of the anonymous letter, with the exception of the guard M____. In this case, it was not possible to corroborate the allegations, but undercover did establish conclusively that the guard was guilty of dereliction of duty. He regularly allowed his friends to exit without making the required package inspections. Other employees had their packages inspected as required.

No satisfactory explanation has ever been found for the letter. A number of theories have been put forth. One of them is that the letter was a deliberate and clever attempt to send the security department chasing after shadows.

From the Public

Strange as it may seem, there are still communities throughout the country today whose residents will occasionally come forward openly and are not afraid to "become involved." An example of the form such an open approach may take is presented in the following memorandum of a personal conversation with an ordinary citizen who gave information only out of a sense of what was the right thing to do.

> Mr. W____ C____:
>
> At your request, I am making this written record of a situation that occurred yesterday, Monday, June 7, 1971.
>
> About 8:30 a.m., Mrs. P____ M____ telephoned me and asked if I would stop by her office at the first convenient time while I was uptown at the Commercial Bank. Mrs. M____ does the office work for W____ Electric Co., this business is located next to the Commercial Bank parking lot. Upon leaving the Commercial Bank yesterday about 11:30 a.m., I went to see Mrs. M____. She related this to me: A friend of hers, a man who lives in A____, who does not work for (the company), told her that (company) distribution dept. (warehouse) employees put cases of shirts in his garage while he was away from home. When he returned home and found that cases of shirts had been placed in his garage, he immediately got in touch with a person or persons and the cases of shirts were removed from his garage.
>
> Mrs. M____ said she did not know how (the company) could stay in business with stealing such as this going on.
>
> I did not press Mrs. M____ for any names and expressed appreciation for the information.
>
> This information was given to Mr. E.B. B____ and Mr. A.M. R____ orally yesterday.
>
> J.M.S.

Many security people will immediately wonder why the executive who conversed with the citizen did not attempt to obtain more details. However, the mere fact that the conversation was reduced to a memo and ultimately passed along to the corporate security department certainly made the whole effort worthwhile. The observation related by Mrs. M____ was duly noted and catalogued for future reference.

Inasmuch as the information dovetailed with other data which had been received during the course of a major investigation, no immediate attempt was made to reinterview the woman. Through other means, the owner of the garage was identified. Ultimately he was interrogated and found to be involved on the periphery of a conspiracy to transport stolen property into interstate commerce. The garage owner, being involved to some extent, was used as a witness in federal court to assist in obtaining convictions.

The group involved had been transporting stolen shirts from Southern Alabama to both New Orleans and Grand Isle, Louisiana, where workers on an off-shore oil rig participated in local distribution efforts. An analysis of the operation revealed that on only one occasion in a period of months did the gang members become so crowded for space in which to store stolen property that they found it necessary to utilize the garage in question. It was purely coincidence that the garage owner happened to be out of town at the time; had he been present, he would have given his permission readily for the use of the garage.

From Other Affected Departments

Experience has shown that the department within a company that is most likely to raise a warning signal as to internal theft is the sales organization. In any volume theft operation, stolen items inevitably find their way into ordinary retail channels. These retail outlets are normally anything but prestige stores—in fact, they tend to be on the low end of the retail spectrum. Nevertheless, once the stolen goods are placed in a retail location, that store becomes a competitor of the company's principal prestige customers in that particular market area.

Stolen goods are usually underpriced in relation to legitimate merchandise, and it does not take long for normal customers to become aware of the fact that they not only have a new competitor for the line, but they are being undersold as well. Invariably, the first representative of the company to receive such a complaint will be the territory salesman. Prudent security executives will make the effort to establish a line of communication with their sales organization so that complaints of this type can filter back to the security department.

Occasionally, an investigation utilizing ordinary shopping techniques will discover that the low-priced merchandise was *not* stolen from within the company, nor was it hijacked. Rather it is off-line merchandise which has found its way into a restricted market area, contrary to company policy.

By the same token, shopping investigations will often identify the merchandise as part of an interstate shipment reported stolen by a common carrier. This situation, of course, is not of primary concern to the security manager but rather is a case which can be passed along to the Federal Bureau of Investigation. In summary, all complaints of the above type received by territory salesmen should be looked into and an appropriate investigation conducted if warranted.

Sales departments are not the only segment of a company which may have occasion to stumble across outside information indicating internal thefts. This may even include the local plant security force, as evidenced in the following inter-office memorandum.

INTER-OFFICE MEMO

FROM: N____ B____ DATE: 6/8/71

TO: W____ C____

SUBJECT: Alleged unauthorized sale of shirts.

About two months ago H____, our security guard, told me that his nephew, W____, a farmer located a few miles south of ______, told him that he had been approached by H____ W____, son of the owner (T____ W____) of W____'s Grocery and Service Station—allegedly a "joint" which is frequented by some of our male employees, principally out of the warehouse—beer drinking, gambling, etc.—who offered to sell him (W____) some shirts which he had gotten from R____ R____, an em-

ployee of our warehouse. He claimed to have recently bought 40 shirts from R____ for $40.00. W____ didn't buy any.
Self-typed
No copies

E.B.B.

In analyzing the above memorandum, it is obvious that the executive who received the information simply sat on it and did nothing until his memory and sense of responsibility were apparently prodded by the receipt of other information, such as the phone call from a neighbor who had witnessed the garage storage incident.

From Employees

Many employees, believing that management does care about theft, will provide solid indicators if given an opportunity to communicate same. Here, whenever possible, the security manager should attempt to have these reports reduced to writing while details are still fresh in the minds of the witnesses. The following two hand-written reports (reproduced in printed form) are examples of the type of information that can be gathered through employees.

6-9-71

Dear Bill,

Around the end of May, A____ D____, one of the clerks at the (company) Outlet Store, came to me to report that on two different occasions over the past several months two people had reported to him that they had been approached about buying dress shirts. The prices quoted were $1.50 each or 80¢ each if bought in lots of 100 or more. He was sworn to secrecy not to reveal the names of the persons reporting this to him. They both declared they would deny it if they were asked about the situation. It was indicated they were afraid of physical harm if their names were involved.

Chris,

D____ and I checked and double checked our shipment of shirts. We found another box (case) shipped to this store and as before this one was completely empty. It contained only the eight small shipping blue boxes. We (D____ and I) counted our shirts individually twice and we both added them. We are exactly 121 shirts short. We came up with a total of 584 shirts out of 705. I included the 12 you carried back to A____. Notice out of 48 short-sleeve shirts we have only 36. Will see you next week.

D____ P____
D____ H____

Received
June 7 1971

All of the foregoing hand-written notes and memorandums-to-file could be replaced with a formal recording procedure utilizing an incident report form, such as shown in Figure 3-1.

In an effort to motivate rank-and-file employees to come forth not only with indicators of theft but with specific allegations, many corporations today have established reward programs for information received from employees. Normally, these reward offerings for internal theft are included in an overall program dealing with security. The subject of internal theft can be "cushioned" by including it with other security matters, as in the extract of a reward program from a major corporation shown in Figure 3-2.

From Paid and Non-Paid Informants

As any experienced security manager knows, dealing with informants is a risky, uncertain business at best. In evaluating tips received in this way, it is always necessary to ask what motivates an informant to give information. To analyze an informant's motives properly can mean the difference between spending a lot of unnecessary effort and being able to zero in on that portion of information which appears to be logical in nature. It has been the author's experience that infor-

INCIDENT REPORT FORM

(For use in all company locations except retail stores)

WHAT:
Theft
- [ ] Employee
- [ ] Suspected Employee
- [ ] From Employee
- [ ] Checks
- [ ] Embezzlement
- [ ] Burglary/Robbery
- [ ] Attempted

Other
- [ ] ____________

Shortages
- [ ] Transfer
- [ ] Receiving
- [ ] Delivery
- [ ] Money
- [ ] Myst. Disap.

Safety
- [ ] Accident
- [ ] Fire
- [ ] Glass

WHEN:
Date of report ____________
Date of incident ____________

WHERE:
Company ____________
Location ____________
City ____________
Phone

WHO
Name
Address Apt
City State

Color	Height	Weight	Build	Hair	Eyes	Comp

Dob Pob Age Sex
Emp
Address
City State
Position Length
SSN
Previous record

HOW (from the facts available describe fully HOW the incident happened or the method used. Time, place in bldg., etc.)

Dept	Description of Property	Value
	Total	

REMARKS: (include any theory or supposition not revealed by facts. Also, names of witnesses or any other information which may prove helpful.)

Distribution
- [ ] Local Mgr.
- [ ] Plant Sec. Official
- [ ] Regional Sec. Officer
- [ ] Corporate Sec. office
- [ ] Other ____________
- [ ] Other ____________

Disposition
Suspended ____________
Ret. to duty ____________
Discharged ____________
Prosecuted ____________
Other ____________

Evidence at:
- [ ] Security Office
- [ ] P. Dept.
Evidence # ____________
- [ ] Returned to stock
- [ ] Other

Report prepared by ____________

Title ____________

CLP 3(8–73)

Figure 3-1. Incident report form.

The Retail Stores Division of
has established a
Loss Prevention Reward Program.

This provides you with an opportunity to help reduce our inventory shrinkage and share in substantial rewards. You can participate through awareness of the following:

Shoplifting:
Employees, other than management, assisting a person authorized to make store apprehensions are eligible for a reward when a shoplifter is apprehended and/or merchandise is recovered. (When more than one employee is involved, the reward will be pro-rated. See back of pamphlet for reward schedules.)

Charge Plates:
When you recover a company charge plate listed on the "Want List", you will receive $15.00. When circumstances warrant, (theft, fraud, etc.) an additional sum may be paid.

Errors:
When you provide information regarding a consistent pattern of errors resulting in loss of inventory and/or money, you will receive a reward.

Internal Theft:
While you are responsible for being alert to the security aspects of your specific job, there are many instances when extra awareness will result in the recovery of merchandise and/or money. When you supply such information, you are eligible for a reward.

Shoplifting and Charge Plate rewards will be paid you directly by your participating employer. In reporting Errors and/or Internal Theft, give the information directly to your company security officer, or if you wish, include all information in a letter, send via U.S. mail to:

Loss Prevention Reward Program
Retail Stores Division

If you wish to remain completely anonymous . . .

This is how the last page of your letter should appear after you have printed or typed the confidential information.

DO NOT SIGN YOUR NAME... but sign with any six digit number of your own choice. Tear off and keep an uneven corner of the last page, bearing the same number.

(choose your own number)
123456.

TEAR HERE

123456

Should you prefer to submit information by phone, call collect (312) 922-
Have all information at hand and your six digit number for identification.

After investigation, if the information you provided results in an apprehension and the recovery of merchandise and/or money, your six digit number will appear either in pay envelopes or a company publication.

When this occurs, mail your part of the letter to the same address for your award. If you wish to remain anonymous, use the return address of a friend, relative or other disinterested third party.

"Remember, any loss however slight, drastically affects profit, your pay and working conditions."

Figure 3-2. Example of a loss prevention reward program.

mation relayed by the average informant is, at best, only about fifty per cent accurate.

Informants can act out of a wide variety of motives. Their motivations can range from hatred and vendettas against other employees or other persons, to jealousy, a desire to become ingratiated with key company officials, a simple desire for recognition, or even a hatred of law enforcement or security persons as evidenced by the occasional trap which is set up by an informant.

Prudent practice dictates that, whenever an investigator is meeting and dealing with an unknown or new informant, a backup security person should be utilized. This is especially true in meeting unknown female informants for the first time, or meeting any informants at night or in strange, out-of-the-way places.

Often, the security manager will be able to develop an informant within the company itself, especially when an occasion presents itself to apprehend some minor offender from among the employees. If handled properly, the offender can often be "turned" and then utilized over a period of time for valuable information.

The security manager must recognize, however, that the pressures involved in the "turning" are not long-lasting, as a rule, and must be supplanted by monetary considerations. In law enforcement, offenders are often "turned" with the implied threat of prosecution. This legal possibility has a lasting effect and most law enforcement persons can obtain a great deal of mileage from such a situation. The security manager, however, is not able to keep the specter of prosecution hanging over an offender's head. He must either prosecute on the offense in question or forget about it if it is allowed to languish. Thus, in order to maintain a reliable arrangement with the informant, a regular payment schedule of some type must be instituted.

It may surprise some to learn that many employee-informants will accept a personal check in payment for their information. These payments can best be handled by the security manager using his expense reports rather than drawing cash funds from the company. A suggested procedure to be utilized on the expense report, so as not to identify the informant, would be to show only the amount, with an explanation, "Payment to informant for information received. Details in personal file." On the security manager's carbon copy of the expense report, which is retained by him, he can then list the name of the informant and whether a check or other written instrument was used. The cancelled check may then be attached to the expense report copy as a supporting document. This procedure, if agreeable to the company, should also be acceptable to the Internal Revenue Service.

The following two hand-written reports (reproduced in printed form) serve as an example of how a minor employee offender can be "turned" into a paid informant. Instructions to the informant were that he was to report once or twice a week on any irregularities or wrongdoing within the facility. For these reports, he was paid a stated sum every week.

Wednesday Night
Oct. 27-1971

Dear Sir,

Reason for no letters lately is because of nothing new to report to you. Things that I know of that has been happening lately is only small things. On Monday Oct. 25- J____ S____ had on a short sleeve shirt and he was cold, so he found him a long sleeve and put it on.

I didn't see J____ L____ take no shirts on Wed-27- but R____ H____ said he saw him take 10- of them. J____ was telling me to day that he had to give a boy 10 shirts to get him to keep a 100- of them for him. Part of the 300- that him and B.K. was going to sell to your boy. Can tell you more about all of that when I see you. And that is about all I have for now.

Wed. 11-3-71

Dear Sir,

Today, Wed,-3-71- worked 2 hours over time, H____ L____ and I were sent to warehouse 7- to pull stock. H____ said that he needed to get him some long sleeve shirts for him to ware this winter. So, I played along with him and told him that I needed some myself. So we got 9 shirts a piece and took them to his' house during the time we were working over time.

Also, A____ J____ got a small box full of shirts. and put them in his car. But didn't get a chance to get a count of them.

And also J____ S____ took him a shirt home with him.

Thats about all of this time

As events turned out, the informant was eventually requested to assist an organized gang of shirt thieves. He was to help steal an illicit order for ten thousand shirts which were to be shipped from Southern Alabama to Atlanta, Georgia. Ultimately, the informant's testimony was instrumental in obtaining convictions in federal court for conspiracy to transport stolen goods into interstate commerce. In cases of this type, the security manager must decide for himself what moral responsibility he has for the safety of his informant after bringing him into open court.

In summary, then, the security executive should be receptive to information from any and all sources. Sources which need cultivation should be cultivated. Other sources which require financial reward should be rewarded accordingly. Some information will be totally false and some information will be close to one hundred per cent truthful, but the majority of information received will fall somewhere in between these two extremes. This middle area will test the logic and ingenuity of the recipient.

SECTION II

THE INVESTIGATION

Chapter 4

Inside Undercover Agents

Having established that internal theft exists, only a naive management will attempt to solve the problem through deterrents and controls.

An enlightened executive group will correctly conclude that hard-core thievery is part of today's business environment. Accordingly, they will realize that identification of the offenders and their separation from the payroll are mandatory. Controls and deterrents are for those basically honest employees who remain.

Management should leave the methods and techniques of investigation to a professional. The security manager is ultimately held responsible for the successful conclusion of any case, and should therefore be able to employ whatever legal investigative methods are indicated.

One of the oldest investigative approaches, second only to interrogation, is the use of undercover agents. Undercover investigation can be defined as a deep as possible penetration of the enemy (or opposition) and the gathering and evaluation of the information which then naturally flows. Many investigations today would be impossible without undercover, and any modern security operation should include it as a vital weapon in the war on internal theft.

SELECTING THE INDIVIDUAL FOR THE JOB

Among the several elements necessary for a successful undercover investigation is selection of the right type of individual for the correct job. Not only should the security manager give consideration to any particular skills which are necessary in the undercover role, but even more importantly, he should be sure that the agent selected is able to blend in successfully with the work force.

For instance, in eastern Pennsylvania, sometimes referred to as "Pennsylvania Dutch Country," it would be futile to select any agent except someone of Pennsylvania Dutch background. By the same token, in a case where the majority of the workers are Chicanos or Mexican-Americans, not only should the agent be able to speak Spanish but he should also be of that particular ethnic background. Work forces comprised of minority groups virtually dictate the use of an investigator from a similar minority group. Because of some of the ethnic background requirements of certain undercover operations, the security manager may find it necessary to attempt his recruiting in certain limited geographical areas of the country in order to insure the hiring of a suitable agent. One advantage in dealing with a local agency, of course, is that such an agency can usually provide agents from the ethnic groups which are indigenous to the area.

Basic Qualifications

Regardless of ethnic or racial background, the good undercover agent usually possesses certain basic characteristics which are worthy of mention. He should be a likable sort of person, possessing a pleasing personality, who basically likes people. He must possess a keen understanding of human nature. He will be perceptive, intelligent, and will possess an above-average memory. The good agent will also have the ability to be successful at inductive reasoning.

Although it might seem from outward appearances that a good agent is anything but honest and loyal, these are two absolutely essential traits. Absolute honesty and loyalty to his employers, supervisors and to himself are necessary. These traits usually speak for themselves when, on the witness stand, the agent creates the impression of a completely credible witness.

Polygraph Examination

Unfortunately, in view of the qualifications stressed above, a surprisingly high percentage of undercover agents engaged in investigative work today in business and industry are *not* basically honest people. Experience has shown that where polygraph screening is utilized in the selection of agent applicants, a very high percentage of rejection is encountered. It is actually shocking to the uninformed to learn how many present-day agents have engaged in impermissible thievery, have retained and utilized drugs which have been secured during an investigation, or who have periodically falsified reports and incriminated otherwise innocent employees.

Realistically, the selection of agents without these undesirable traits can only be accomplished through a good pre-employment polygraph examination.[8] In those few states where polygraph screening is outlawed, the next best substitute would be a written honesty test such as the Reid Report.

Inasmuch as most undercover operations are somewhat unrestrictive as to the behavior of the agent while on assignment, it is necessary to place some safeguards for the protection of both management and employees as well as a control on the agent himself. This can best be accomplished by periodic polygraph examinations. Where these are utilized, the following four questions should be included:

1. Did you keep an accurate account of all company mer-

---

[8]For a list of polygraph practitioners, the reader is referred to: Secretary, American Polygraph Association, Wayne M. Boysen, University of Minnesota, 2030 University Ave., S.E., Minneapolis, Minn. 55455.

chandise you took on your last undercover assignment?
2. Did you pay for all company property which you took and did not turn in on your last assignment?
3. Did you falsify any of your undercover reports on your last assignment?
4. Did you deliberately fail to report any significant security violations on your last assignment?

In-House vs. Agency Investigators

Along with the foregoing selection process, security managers of large corporations are occasionally tempted to try an in-house approach to undercover work. Many such executives have recruited graduating college seniors with degrees in industrial security, police science or some related field. Experience has shown that there are pros and cons on both sides of the in-house vs. agency argument. Some of the differences which security executives must be alert to are highlighted in the following comparison:

AGENCY OPERATIVE	POLICE SCIENCE GRADUATE
1. Usually good at "roping." (See explanation below.)	1. May be poor at "roping."
2. May not know legal aspects of the work.	2. Usually has insight into legal considerations.
3. May be a poor report writer, necessitating a rewrite man.	3. Can usually write acceptable reports.
4. Sometimes makes a poor to fair impression on judge or jury.	4. Usually comes across as a professional in court.
5. Usually blends in and is "street wise."	5. May not blend in and is "straight."

The above comparisons are certainly not meant to be absolute, but rather are factors which occur at a rate greater than chance. Most deficiencies can be corrected through adequate training; even the lack of a formal college education can be offset by training and experience. Although a large company may be able to hire police science graduates, without question the only sensible route for the smaller company is to utilize agency operatives. Several large corporate security departments successfully utilize both.

Cover Stories

Cover stories are an extremely important part of any undercover operation. The more complex the case and the greater the sophistication of the criminal suspects, the more attention must be given to the preparation of the cover story. An agent can only be as effective as his cover story is acceptable. It must fit the role being played by the agent. Many authorities on undercover stress that there are three rules to observe on a cover story:

1. Keep it simple.
2. Keep it believable.
3. Keep it as close to the truth as possible.

SUPERVISION AND COACHING

The amount of previous experience possessed by an undercover agent will determine the amount of coaching and direction which he should be given by his supervisor. Many experienced agents are able to virtually take a case and work it singlehanded, with minimal or no supervision. The beginning agent, of course, will require the most supervision and direction, and the security manager must recognize this.

If the security manager is dealing through an agency rather than an in-house approach, he may conclude that he wants to participate in the direction and coaching of the agent if there is a certain lack of experience. On the other hand, the security manager may have complete confidence in the agency and will be content to defer coaching and direction to agency supervision.

There is no set formula as to the proper approach to coaching and direction. In some cases, several routine telephone conversations during the week, followed by a monthly meeting, may be sufficient. On the other hand, the security executive may well decide that the case is sufficiently complex to require meetings as frequently as once a week. A word of caution, however, is in order regarding the selection of a meeting site. Months and months of good, productive undercover investigation can be undone by the chance encounter of the agent and his supervisor with other employees of the plant in some restaurant, bar, or other public place. Although there is a tendency, on occasion, for the agent and his supervisor to desire a get-together for dinner, it is best that any meetings be restricted to a motel room and preferably be held after dark.

"Roping"

The art of roping can be defined as the undercover agent's gaining the trust and confidence of a suspect to the point where the suspect will disclose past or current criminal acts, or at least will not hide them from the agent's view. Techniques used to accomplish roping need only steer clear of entrapment and areas of illegality. Roping suggests the old adage, "It takes a thief to catch a thief." The theory here is that an established, mature thief would never reveal himself to a newcomer (the undercover agent) until he feels that he (the agent) can be trusted or at least is not a threat. Therefore, if the agent can create the impression that he, too, is dishonest, he will have neutralized some of the defenses of the older thieves.

For all but the most experienced agent, roping is the one area that seems to require more push and direction on the part of the supervisor than any other. In this legally sensitive aspect of undercover operations, the agent must be prodded when necessary, but restrained when it would appear that his roping activities stray across the border into the area of entrapment.

Balance Between Two Employee Groups

Another area of coaching required by the inexperienced agent is the ability to avoid becoming identified too closely with any particular plant group. In any industrial setting one can usually identify two main groups of employees—those who are relatively "straight" and free from dishonesty, and those who care little about the company and take advantage of every opportunity to further their own ends either through company expense or outright illegality.

In the initial stages of an undercover investigation, it is often necessary for the agent to gain and keep the good will of the "straights." Members of this group often possess invaluable information about the illegal activities of persons in the other group. During this time the agent, in effect, must walk a tightrope between the two groups so as not to be identified too closely with either one to the detriment of his future relationship with the other. Eventually, the agent will have to decide to swing away from the "straight" group in order to become completely accepted by the thieves in the plant. This switching over from one group to the other is accomplished through the art of roping. It usually results in the alienation of the original group, which is loyal to the company and not interested in stealing.

Making "Buys"

Many agents must be coached not only in the art of making "buys" but also when to buy and what to buy. Newer agents often have a tendency to attempt "buys" simply for their own sake. No agent should ever attempt to buy contraband that he could easily obtain through his own efforts at theft. In other words, if the plant has a certain commodity which would be impossible for the agent to obtain through his own theft activities, then a "buy" would be logical. If the agent could obtain the item by stealing it himself, a "buy" would be completely illogical. It is sometimes easy to lose one's perspective when personally engaged in undercover work, and for this reason coaching on all of these points is highly desirable.

Outside Agents

The supervisor of the undercover case must also be the one to make the decision to introduce an outside agent. Many times, the inside undercover agent will be able to identify certain thieves among the work force but will not be in a position to further his investigation on the outside. Outside investigation can include "tail jobs" at quitting time and surveillances of subjects' homes. Obviously, these activities should never be attempted by the undercover agent because of the risk involved; they should be performed only by outside agents.

The director of the investigation is in the best position to determine when an outside agent should be added to the case. He must also determine the best method of communication between the undercover agent and the street agent for purposes of tailing. Usually, the inside undercover agent will be in a position to "finger" a particular suspect to the street agent as a suspect leaves the plant at either lunch time or quitting time.

Maintaining Perspective

The last (and most important) requirement for the undercover supervisor is that he maintain an overall perspective on the entire case. The typical undercover case will involve the identification of various dishonest employees on the work force. This is usually done employee by employee. In other words, over a given period of time, the undercover agent will gradually be able to identify more and more of the dishonest employees on the company payroll. It is up to his supervisor to insure that he does not become bogged down in his relationship with any one dishonest employee or group of thieving employees but endeavors to proceed and make new identifications, even while maintaining contact with dishonest employees previously identified. By keeping the "big picture" in mind, the supervisor sees to it that the case is developed to its logical conclusion so that the majority of dishonest employees in the plant have been sufficiently identified to warrant interrogation.

COMMUNICATION WITH MANAGEMENT

It is virtually impossible to visualize any type of situation arising in an undercover investigation wherein it would be necessary for the agent to make direct contact with management regarding a progress report on the case. In some limited situations, it may be permissible for the agent to query management for additional information that would be beneficial to him in his pursuit of the case.

Avoiding Premature Exposure

Occasionally, there is a tendency on the part of the inexperienced agent to "push the panic button" and attempt to make direct contact with management when he learns of some serious act of dishonesty which is about to take place. Rather than notify management of such an impending situation—which management is ill-equipped to handle anyway—the agent should be content to allow the situation to proceed on its normal course and simply take full advantage of every opportunity to gain information about the operation. To precipitate an overt act on the part of management in attempting to thwart a major theft, the agent would only insure the premature exposure of the whole investigation and probably would preclude the eventual identification of additional thieves.

Limited Progress Reports

Members of management should be made aware of internal undercover investigation on a "need-to-know" basis only. The reason for this is not that management is to be mistrusted, but by furnishing management with progress reports the security manager places them in the position of having to be expert actors, which they probably are not. No longer can management deal effectively and in a routine manner with employees who they now know have been identified as major perpetrators of theft. Even with coaching to the contrary, management has been placed in the position of having to "put on an act" in order to appear normal. This unnatural behavior can easily be detected by the rank-and-file employees.

Furthermore, there is a natural tendency on the part of management to attempt to correct or thwart bad situations or illegal acts. In attempting to stifle such an urge, once again the management person is put in the position of playing an unnatural role.

If progress reports are to be made, they should be in the form of verbal briefings to senior management only, rather than to operating management. These reports should be general in nature and should simply cover the general progress of the case to date.

One reason for excluding operating management from progress reports is that the security manager will not want any tightening of internal controls during the undercover phase of the case. Rather, it is often desirable to have management slack off on their enforcement of their internal house rules so as to give the thieving group more opportunity for dishonest activities and to make it easier for the undercover investigator to do his job.

The ideal time for a more complete progress report would be on the eve of a major "bust," following the undercover phase of the investigation. In this way, management will feel that they are being included in the security efforts but will not be in a position to compromise or otherwise stifle the investigation, which has now been completed through the undercover efforts.

It cannot be stressed too forcefully that the most successful undercover cases have always operated on a "need-to-know" basis. The security agency supervisor obviously does not have as much ability to restrict this information as does the corporate security executive. The agency supervisor who presides over the client-agency relationship involving undercover agents can only counsel and attempt to educate the operating executives of his client as to the adverse possibilities of premature actions on their part which could also result in serious consequences to the undercover agent himself.

Regardless of the approach, however, the fact remains that the majority of industrial theft cases today are being broken by undercover investigation. The security executive who avoids the use of undercover when it is clearly called for is simply failing to meet his greatest challenge head on.

Chapter 5

Outside Undercover Agents

INFILTRATION

Outside undercover agents are normally utilized to complement the efforts of an inside agent or, in some cases where inside agents are not possible, to handle the investigation entirely from the "street." In the industrial setting, the "buy" may be the end product of the efforts of an outside undercover investigator, but basically his normal objective would be the infiltration of designated places and/or certain groups of people.

Cover Stories

Unlike the elaborate precautions and efforts directed at setting up a cover for the inside undercover agent, the external or outside undercover agent can often get by with a cover story that will withstand only a cursory inspection. Motto, in his book *Undercover,*[9] deals extensively with how to set up cover stories for short-term, outside undercover assignments and making "buys."

---

[9]Motto, Carmine J., *Undercover.* (Springfield, Illinois: Charles C. Thomas, 1971).

An individual cover story for an outside undercover agent, like that for an inside agent, has to be designed to take into account any complexities of the case and, of course, must be in keeping with the projected time span of the investigation. In other words, the more serious the case, the more effort will generally be put into establishing an individual cover. The most difficult cover stories will be those which must be constructed for "street" agents who plan to work in small rural communities where everyone is either known to or at least generally recognized by most of the local residents. The following extract of a case history offers an example of how such a group cover story can be put together.

Bartow County, Georgia . . . . After an initial inside undercover investigation had proved successful, it was decided to supplement the inside agent's efforts with a team of outside undercover investigators. The purpose of the outside team was primarily that of surveillance of various M.O.'s, or methods of operation, which the inside agent had discovered were being utilized to remove stolen merchandise.

The team, consisting of two males and one female, headquartered at a local motel which catered to construction crews for the power company and the railroad. The cover was simple and to the point—the team claimed to be from a survey company engaged in environmental studies. Specifically, they were supposedly making an air pollution study of the area in relation to the numerous fabric and textile mills which are located in the county. This story was augmented by appropriate signs on the surveillance van, company name tags, and safety hats with company insignia. The surveillance was programmed to run for a period of six weeks and was successful in a community that normally would be somewhat suspicious of out-of-town strangers.

Another successful group cover story has been an engineering survey team, complete with some of the equipment "props" normally associated with engineering field survey work. As in all undercover work the possibilities of adequate cover stories to fit the particular situation are only limited by one's imagination.

Infiltration Leading Up to the "Buy"

As an example of how certain places and groups of people may be infiltrated, the following case history reveals the events and efforts leading up to the "buy" which was described in the Introduction.

Covington County, Alabama . . . . Through the efforts of one outside agent who was working undercover as a local taxicab driver, investigators were able to pinpoint a local private club as a hang-out for most of the criminal element in the community. It was felt that, by infiltrating such a place, an agent could pick up a trail in connection with stolen shirts. A second agent was therefore assigned the task of penetrating the club.

On his first several visits he was able to gain admission but was unable to strike up any conversations of note and generally was received coolly. On both of his early visits to the club, the agent noticed a young lady patron whom he had previously observed working in the community. He made it a point to meet this young lady outside of the environs of the club and make her acquaintance. After a short time, the two began dating. At the young lady's suggestion, they began spending a number of their evenings at the club, which was her favorite spot for social activities.

The agent had previously observed that the young lady was very friendly with the bartender, the bouncer, and other regulars of the establishment. Through her, he became friendly with the club's bouncer and was eventually accepted as a "regular" himself. The bouncer ultimately became the pistol-packing "salesman" who took part in the conspiracy to sell ten thousand shirts at a meeting at the Atlanta Airport. He was not an employee of the company, but through his fencing activities on behalf of the employee theft ring, the direct link was made.

Transportation

In any infiltration of a small rural community, transportation can be a problem. Experience has shown that the use of late model rental cars is completely counterproductive, as these generally tend to be associated in local residents' minds with the type of transportation used by federal agents.

If a moving surveillance is to be conducted, then the investigator should utilize slightly older model automobiles that are in excellent mechanical condition. If moving surveillance is not a critical point in the case and the vehicle is needed to provide transportation only, then even older cars can be procured, as long as they are basically in keeping with the cover story which has been set up. Registration of the vehicles can either be in the names of the individual agents or in the name of the "cover" company, e.g., "Atwell Survey Co." Insurance coverage can usually be accomplished by the corporate insurance department's simply adding such cars to the listing furnished to the insurance carrier. If the corporation is self-insured, then independent liability insurance should be purchased.

With the advent of the CB radio boom in the United States, the appearance of a whip antenna with magnetic base on the roof top of an older model car raises little or no suspicion. Furthermore, it enables the investigators to conduct their surveillance using their own security radios, which will normally be in the VHF or UHF ranges.

MAKING THE "BUY"

A "buy" in the corporate security field may encompass two purposes:

1. To obtain the physical evidence of the crime.
2. To enable the investigators to determine the actual method utilized in removing the goods.

Many times, because of his isolated work assignment, an inside undercover investigator is unable to discover the actual M.O. which is used by the thieves. In order to gain this information, he has to depend on an order for a "buy" to generate enough activity so that the street agent can observe the method utilized.

Many times, the "buy" can be initiated by the inside undercover agent himself. That is, the inside agent can offer to introduce company thieves, whom he has already identified, to an "acquaintance" who is in the market for stolen goods. This offer cannot be construed as entrapment if the agent has already been able to establish a pattern of theft. The courts have generally held that such an offer is merely giving the thieves an additional opportunity to carry forth a scheme which had already been in existence. As pointed out earlier, the inside agent should never attempt to make a "buy" of something that he himself could steal as easily as the thieves.

Photographing and Recording

If the agent is successful in interesting the thieves in discussing a "buy" with his "acquaintance," every effort should be made, consistent with legality, to obtain evidence of such a meeting. Motion pictures or still photos may be taken of the participants of the meeting as they arrive at a designated place. Actual "bugging" of the room in question or even of one of the undercover participants is also desirable. Unless a particular state law forbids all electronic surveillance, the bugging of a room or "wiring up" an agent does not violate the law, as long as it is the agent's room and/or the agent is a party to the conversation with the suspect. Tape recordings of such a meeting are often valuable in court if the meeting is handled properly and the recording itself is of good quality.

Such meetings should usually be held on apparently "neutral" ground, and in places that would minimize the chances for injury to the participants and maximize the opportunities for photographic or recording coverage. Motel rooms usually prove ideal for this type of activity.

Figure 5-1. Photograph taken by an investigator just after a "buy" has been made. Goods have already been transferred to trunk of courier's car.

Figure 5-2. Photograph taken through windshield of surveillance vehicle shows participants in a "buy" haggling over price.

Negotiating With the Thieves

The second agent, posing as the fence, should be an excellent actor and be able to negotiate with the thieves much as any merchant would who was engaged in buying stolen merchandise. If the agent seems too eager to buy, the whole deal can be jeopardized. For this reason, he should appear to be reluctant to part with his money unless there is absolutely no question of getting adequate value in return.

Agents entering into such negotiations should be well versed in the true value of the items for which they are negotiating. Generally speaking, no more than 25 per cent of the indicated retail price should ever be paid for stolen goods. Actually, the price paid can be an even lower percentage, depending on the quantity of goods to be ordered and also whether there are any unusual circumstances which would put the thieves in a position of wanting to make a very fast sale. The agent making the "buy" should not hesitate to haggle over minor details as any dishonest merchant would be likely to do. For instance, a discussion about whether or not labels have to be destroyed can often contribute to the overall roping effort, and in some cases may also contribute to a reduction in price if the labels are to be removed by the buyer.

Before the meeting takes place, the buying agent should also insist that the thieves produce samples of the product which they intend to sell at the meeting. If such samples are not forthcoming, the buyer should insist upon delivery of the samples prior to any deal being consummated and the money changing hands.

Eliminating the "Middle Man"

One favorite tactic of many security chiefs is to have the second agent, posing as the buyer, attempt to eliminate the inside undercover agent from any further transactions. Generally, this is easily accomplished by the second agent's simply saying to the thieves, "Why don't we deal direct—we would all make more money by letting John out of it." Most thieves will jump at the chance to eliminate a "middle man" and save any "cut" which they feel they might have to pass along. They usually explain this away by later telling the inside agent that negotiations have fallen through and that no deal is possible. They assume, of course, that the buyer will hold up his end of the story should the inside agent check back as to its truthfulness.

By severing the inside agent from further transactions, the risk of losing the case through charges of entrapment is greatly reduced. The whole idea is to attempt to be able to show that the actual stealing of the goods by the thieves is not being influenced by the inside undercover agent, and that the "buy" itself represents nothing more than an additional outlet for the thieves to make their sale.

Precautions Against Robbery

As in law enforcement cases, great caution must be used when the actual "buy" itself takes place. The thieves know that the buyer or his "courier" or pick-up man will usually be carrying a large amount of cash. It is not uncommon to encounter thieves with such shortsightedness that they attempt to rob the pick-up man of the money and deliver absolutely no goods whatsoever. This can sometimes be circumvented by having the "buy" itself covered by ample undercover agents in the vicinity, whose very presence tends to discourage an armed robbery but would not be counterproductive insofar as the actual "buy" is concerned. On some occasions, where an inside undercover agent has been close to the theft group, it is possible to learn of holdup plans beforehand and to take adequate precaution.

In the event that a holdup does take place, the pick-up man should be instructed to cooperate to the fullest and simply hand over the money. With ample witnesses in the locale, it would then be an easy matter to file armed robbery charges against the thieves at the appropriate time. Armed robbery usually carries a much greater penalty than grand larceny and is just as effective a method of separating wrongdoers from a company payroll as any other.

In arranging for the "buy" to take place, common sense will dictate the selection of a location which is in the open, has at least some people in the area, and is in daylight. All of these things may not always be possible but certainly are highly desirable from the standpoint of the security manager.

Another precautionary tactic which has been employed from time to time with success is that of "buying short." In the industrial setting the security manager is utilizing company funds to pay for company merchandise which has already been stolen; naturally, he will want to minimize the expenditure of such funds. The way to do this is for the buying agent to simply give his pick-up man less cash than was agreed upon, with instructions to bring back less goods. After all, the ultimate objective here is simply to secure some of the goods as evidence, and it is reasonable to assume that much of the remaining goods can be recovered through a search warrant or consent to search.

"Buying short" involves a certain risk of alienating the

thieves to the point where the transaction may not be completed. An astute security manager, however, knowing the psychological make-up of the thieves with whom he is dealing, can usually be successful in projecting the outcome: namely, that the thieves will take the position that some money is better than no money and that they have been able to move at least some of their stolen goods. If the "buying short" tactic is used, however, the thieves should be left with the clear impression that more "buys" are to come and ultimately they will be able to sell the original quantity of goods that was discussed.

Transporting Contraband

In planning for the transportation of contraband from a "buy," security personnel should be careful not to utilize personal automobiles if there is a chance that the vehicles would be traceable and would compromise the investigation in some way or endanger the operative. Generally speaking, a rental car or truck which cannot be traced is the most desirable way of picking up contraband and transporting it from the scene to the location where it will be held.

Preservation of Evidence

The rules involving the preservation of evidence obtained through "buys" or any other means are no different for industrial security personnel than for law enforcement personnel. The chain of possession of such evidence must be clearly shown to the satisfaction of any judge or arbitrator. This is simply accomplished by having the pick-up man (using the foregoing discussion as an example) place his initials and the date and time on the cartons or containers as quickly as it is convenient for him to do so without jeopardizing his cover. If the pick-up man then turns the goods over to another security person who has been designated to maintain custody of the contraband, then that agent must also indicate on the container his initials, the time and the date he takes a receipt from the first agent.

The main point is to be able to show who had sole custody of the goods during any given time span, and to prove that no one else was in a position to tamper with or alter the evidence in any way. It is also suggested that the written report of the agent involved in the handling of the contraband should indicate the marking of such evidence for identification. If possible, the nature of the identifying marks and their location on the contraband should be included. In this way the written report stands to corroborate the actual identification mark on the evidence and thus greatly bolsters the entire case.

"Switching" Contraband

A very valuable approach to obtaining physical evidence of theft can sometimes be worked out between an inside undercover agent and a "street" agent who, working as a team, can endeavor to "switch" stolen merchandise. An example of this would be a situation where the inside undercover agent has positively identified a fellow employee as regularly stealing certain items of company property. If the thief in question frequents a tavern or some other location such as a pool hall after work, he may be in the habit of carrying the stolen items with him in a paper bag or some other container. It is sometimes possible to switch paper bags or other similar containers, leaving ones filled with fruit or some other innocuous items in place of those filled with contraband. The stolen merchandise thus seized can later be used as a basis for a specific complaint of larceny on that date. Experience has shown that when such switches are made, the thief usually concludes that the mix-up was the result of a natural mistake.

Seizure of Company Property

On some occasions, it may be possible to "steal back" or seize company property which is in the process of being removed. An example of this would be where certain stolen items are placed in the bed of a pick-up truck parked in the company parking lot. If the opportunity presents itself, an outside "street" agent can sometimes remove one or two of the specific stolen items from the bed of the truck. At the worst, the thief will conclude that part of his load has been "ripped off" by another thieving employee. From the agent's standpoint, he now has gained possession of some actual physical evidence of larceny which can be charged to the suspect.

Marking Contraband

On other occasions where it is not possible to make switches or seizures, either the inside or the outside agent may be in a position to mark the contraband for future identification should it be recovered during a subsequent search. This is most easily accomplished by utilizing a fluorescent crayon which can be obtained from any police supply company. If the crayon is not available, then any identifying mark, even with a ball-point pen, can be made by the undercover agent.

In one case where a systematic stealing was in progress over a period of months and the only desire was to mark the evidence, a special rubber stamp was made up signifying "Nov. 31, 1976." This was utilized to stamp cartons and aroused absolutely no suspicion on the part of the thieves involved in the operation. Of course, such a date does not exist, and therefore the unique stamp became an identification mark that would easily hold up in court.

SHOPPING INVESTIGATIONS

A major form of street type undercover investigations utilized today is called "shopping." Shopping investigators are used primarily to service the retail store industry but may also be utilized in other types of establishments such as restaurants, hotels, bars, etc. For purposes of illustration, however, we shall restrict our discussion to shopping investigations which are directed towards the retail store industry.

Shopping investigations are offered by a number of private agencies around the country who service the retail field. In recent years, a number of larger retail companies have undertaken shopping investigations on an in-house basis.

Service and Integrity Shopping

Basically, shopping investigations can be broken down into two types: service shopping and integrity shopping. The old rationale behind service and integrity shopping was that they provided management with a tool to evaluate the type of service being rendered by the salespeople to the customer (service shopping) as well as an honesty test of the personnel themselves (integrity shopping). Many security chiefs in recent years have come to question the advisability of combining the two types of shopping into one investigation. In increasing numbers, retail security managers have come to the conclusion that integrity shopping is compromised or weakened to some extent when it is combined with service shopping. The reason for this is that the investigator must give so much concentration to the reporting aspects of the service shopping that he is unable to devote sufficient attention to the techniques needed to insure a good integrity "shop." Furthermore, many members of shopping investigation teams are excellent at service shopping but are poor producers when it comes to making integrity "shops."

Integrity Shopping Techniques

Essentially, an integrity "shop" can be defined as creating an atmosphere in which the sales clerk will be likely to commit a dishonest act if he or she has the propensity to do so. Here again, this is not to be confused with entrapment. Integrity shopping gives the thieving employee an opportunity to do the very thing he or she does several or more times a day with legitimate customers when the opportunity presents itself. The techniques utilized by shopping crews are many and varied in their approach, but all have in common the fact that identification of the questionable transaction can be made by so-called tie-down sales to other members of the shopping investigation team which are properly recorded on the cash register or written sales receipt.

A typical integrity shopping crew will consist of two or

preferably three people. The best shopping crews seem to be made up of full-time shoppers who have developed their art and are very proficient at it. An alternative to a full-time shopping crew would be a nucleus of at least one full-time or lead shopper, supplemented by part-time shoppers. Often, part-time shoppers can be recruited from the ranks of housewives who at one time or another have worked as a member of a full-time shopping crew.

The crew is usually directed by a so-called lead shopper or crew chief who is responsible for dispensing of the cash to make the purchases, transporting the team, and working out the routing and schedule. The crew chief will also act as liaison with his or her principal in the agency office or, in the case of in-house investigation, with the retail security manager.

It will normally be up to the agency supervisor or the retail security manager to make arrangements with the crew chief for the storage of the purchases while the particular series of stores are being shopped over a period of days. Ultimately, all of the merchandise is normally turned in to the company for full credit and this is usually handled through a high-ranking member of the controller's staff of the store. In this way, rank and file sales personnel are not aware of shopping investigation merchandise being returned for credit.

Figures 5-3 and 5-4 are examples of some of the report forms which are utilized by either agency or in-house shopping investigators.

SAMPLE SHOPPING INVESTIGATION REPORT

FIRM ______ Store No. ______ Case No. ______

Address ______ City ______ State ______

Date ______ Time ______ am/pm Opr. ______ Report No. ______

NAME
NUMBER
LETTER ______

Sex ______ Age ______

Height ______ Weight ______

Build ______

Eyes ______ Nose ______

Teeth ______

Complexion ______

Hair color ______

How combed ______

Glasses ______

Jewelry ______

Other ______

Reg loc/no: ______

Reg read: ______

Other cust/oprs/salespeople: ______

PAYMENT MADE

	$20	$10	$5	$1	50¢	25¢	10¢	5¢	1¢	
1.Pur										Trans # ___
2.Pur										Trans # ___

DESCRIPTION OF TRANSACTION

SALESPERSONS APPEARANCE:

___ Well groomed
___ Passible
___ Average
___ Unimpressive
___ Unkempt
___ Other

SALESPERSONS ATTITUDE:

___ Enthusiastic
___ Pleasant
___ Routine
___ Indifferent
___ Antagonistic
___ Served promptly
___ Suggested other items
___ Offered a 'thank you'
___ Other

PURCHASES MADE

TOTAL		

Figure 5-3. Sample shopping investigation report.

DATE Feb. 3-9 1978 UNIT	OPR.	START	END	OUT	HOURS	ISSUE	MDSE.	BAL.	SIG.
CASHIER MD	M.D.	600.00			18	200.00	189.92	10.08	MD
END MI.	B.O.				13	200.00	191.50	8.50	BO
BEGIN MI.	C.N.				18	200.00	160.65	39.35	CN
TOTAL MI.									
CAR USED						600.00	542.07	57.93	

$ ON HAND A.M. $ 200	TOTAL ENT.
ADDITIONAL $ 400	
TOTAL START $ 600	TOTAL TESTS 29
	PER
TOTAL MDSE. SPENT $ 542.07	DIEM
BALANCE FORWARD $ 57.93	Itemize Exp. – Reverse Side

Store	M.D.	B.O.	C.N.	Entries	T	Total
Case # Feb 3, 1978	4.65	12.99	8.75	26.39	6	6
Boyd's		3.25	10.89	14.14		
Clayton	18.25	21.79		40.04		
	22.90	3803	1964	80.57		8057
Case # Feb 4, 1978	19.95	8.29		28.24	5	11
Boyd's	16.75	28.76	1684	62.35		
Crestwood	13.99		2250	3649		
	50.69	37.05	39.34	127.08		20765
Case # Feb 5, 1978	6.64		32.55	39.19	4	15
Boyd's	17.92		260	20.52		
Jamestown	21.75			21.75		
	46.31		35.15	81.46		289.11
Case # Feb 7, 1978	17.95	3606	16.22	70.23	6	21
Boyd's	1892	19.99		38.91		
Northwest Plaza		15.50	19.75	35.25		
	36.87	71.55	35.97	144.39		433.50
Case # Feb 8, 1978	8.75	16.95	2.45	28.15	5	26
Boyd's	350	27.92		31.42		
ST. Clair	13.85		16.99	30.84		
	26.10	4487	1944	90.41		523.91

Store				Entries	T	Total
Case # Feb 9, 1978	2.89		7.82	10.71	3	29
Boyd's	4.16			4.16		
West County			3.29	3.29		
	7.05		11.11	18.16		54207
Case #						
Case #						
Case #						

This is a Cash and Production Control form. Properly completed, it gives an accurate record of what has been done in each store. It tells the hours worked by individuals in a crew and an accounting of the shopping money issued, spent and returned.

Figure 5-4. Cash and production control form.

Chapter 6

Tailing and Stationary Surveillance

Moving and stationary surveillance capabilities are required skills for any well-rounded private security operation. Not only are surveillances often necessary to augment inside undercover operations, but often they may represent the only means of building a case against a suspect. Even in the case of street type undercover work where penetration of a certain location or the roping of a particular individual is indicated, surveillance is almost always a first step.

VEHICLE SURVEILLANCE

The ability to conduct successful moving surveillance with today's varied traffic conditions is an art. Good "tail" men must have the ability to perform successfully in all types of conditions, whether it is expressway driving, downtown traffic conditions, interstate highway travel or rural country roads. It becomes readily apparent that only expert drivers can master the art of successful vehicular surveillance. Persons who are knowledgeable in this area would agree that the best tail men are often to be found employed as operatives for private detective agencies which specialize in marital investigations. Some of these expert drivers are successful in maintaining weeks of tailing of male subjects who have been alerted beforehand to the strong possibility that such a surveillance may exist. Nevertheless, in spite of the evasive maneuvers which may be attempted by the subject who anticipates a surveillance, these tail men are more often successful than not.

For the beginning investigator, training in the art of vehicular surveillance is absolutely necessary. In surveillance work the subject may become "burned"; that is, suspicious or aware of the fact that he is under surveillance, without being able to pinpoint any particular agent. When the term "burned" is used in connection with an agent, it is synonymous with being "made" or identified by the subject. In this case, the agent's usefulness on the assignment in question is at an end.

As many evasive tactics as there are to throw off a surveillance, there is an equal number of tactics which can be employed to reduce the chances of "burning" the subject of the tail. As many of these tactics as possible should be demonstrated to the beginner, and from there his own imagination and experience at this type of work will further expand his abilities. Such training is not unlike the defensive driving courses taught by some of the major automobile insurance companies. Because the scope of this text does not permit an extensive discussion of vehicular tactics, we will confine our consideration to general points which must be borne in mind by the director of any surveillance operation.

The Surveillance Vehicle and Equipment

Rental cars are generally considered to be the best choice for vehicle surveillance. For a surveillance involving more than one person per vehicle, a four-door sedan is preferable to a two-door, as the passenger can often utilize the rear seat to enable a different "picture" to be presented to the subject of the surveillance. The vehicle should be chosen carefully so as not to be conspicuous in color. On surveillances running a number of days, cars should be switched with the rental agency on a daily basis if at all possible. Some rental agencies will be amenable to maintaining a weekly rate in spite of the daily changes of automobiles, especially if full-sized sedans are being used.

Items kept in the surveillance vehicle should include changes of outer clothing such as caps, hats and a reversible jacket or topcoat. Surveillance cars should also contain a pair of good quality binoculars as well as a motion picture or still camera equipped with telephoto lens. A closed container for liquids is necessary for relief of personal needs during car surveillances which will run an indefinite number of hours. A newspaper should be present in the vehicle to enable the operative to have cover for his camera work and also to provide a cover for himself when in a parked position. A person sitting in an automobile appearing to read a newspaper draws far less attention than someone who appears to be doing nothing except sitting.

Some agencies of the federal government, along with private agencies which utilize automobiles regularly for tail jobs, have made certain modifications in the automobiles' lights. For instance, by the use of interior switches, either the left or the right headlight can be shut off, or, on the other hand, both lights can be kept turned on. These various combinations, of course, add to the number of different "pictures" which are presented for the subject's rearview mirror. Interior lights should also be modified to the extent of taping down the spring buttons found in the door panels so that the lights do not come on when the doors are opened. A good investigator should never remove the bulb from the interior dome light; the dome light must be kept in operating condition, for reasons explained later in this chapter.

Risks in Automobile Surveillance

In training the new agent, considerable time should be spent in attempting to inculcate in him an understanding of when it is permissible to take chances and when it is not. Any successful automobile surveillance of necessity involves a certain amount of risk to passengers of the car, pedestrians, and other vehicles on the street. These risks are further increased when it becomes necessary to violate traffic laws such as running stop signs and red lights or excessive speeding. On a long-term tailing assignment, it may be highly desirable to minimize the taking of such chances, even though the chances of losing the subject are thereby increased. It would probably be reasonable to state that maximum traffic violations would be permissible only in the event that the agent knew the "buy" had occurred or was about to take place, or that contraband was being transported to a specific location. Only at a time like this can it be justifiably said that it is not affordable to lose the tail.

Multiple-Auto Surveillance

Obviously, the most difficult car tails are those which involve subjects who own powerful automobiles or are high-speed drivers. The high-speed driver is enough of a hazard on the road, and for this reason the surveillance director has to exercise prudent judgment on how a successful tail job can be accomplished. For the successful tailing of such a subject, multiple automobiles must be used—a minimum of two and possibly three or four, all with adequate radio communication. This could involve a strategy of several of the cars leapfrogging with the subject, running parallel streets, or even a combination of both plus other strategies that can be employed. In attempting to determine a regular route of travel of such a high-speed driver, it may be prudent for the director to utilize a number of days of surveillance, forfeiting the loss of the subject's car each day but at the same time gaining additional distance information on the subject's regular route of travel.

For the department or agency which engages in a heavy volume of vehicular surveillances, there are now devices on the market which can be attached beforehand to the subject vehicle to emit a radio signal. This, of course, requires a combination receiver and direction finder in one of the tail vehicles. The small "beeper" device is easily planted on a target vehicle and becomes extremely useful when visual contact with the vehicle has been lost.

Truck Tails

Generally speaking, truck tails are much easier than automobile tails and in corporate security work are probably the more prevalent of the two. Because of their size and unusual markings, trucks are much easier to keep in sight than are automobiles. One thing that the beginning agent must be aware of is the large sideview mirrors that are used with regularity by all truck drivers. For instance, in making a right angle turn, the tail car should always make a much wider turn so as to keep out of view of the particular sideview mirror involved. On the other hand, because sideview mirrors represent the only means of viewing to the rear for the truck driver, there is also an advantage which can be taken by the agent. This advantage lies in a visual dead spot immediately behind the truck, out of view of both of the sideview mirrors. This of course involves tailgating, and a high degree of driver capability is required to avoid an accident. However, in heavy downtown traffic conditions this often represents the only method which can be used to stay with the truck at various intersections and through traffic signals.

In corporate security work, most truck surveillances are of company trucks which are regularly making deliveries or pickups. Accordingly, these trucks generally follow a prescribed route, and it is often possible on any given day to be able to determine beforehand which stops are to be made by the driver. Often the surveillance agent will make a dry run of the route beforehand to learn of traffic conditions, special turns, bridges, turn-offs, etc. In making the dry run, the agent also has the opportunity to spot various locations along the route which he will recognize as being places that can be used in his own tactics to minimize his exposure to a "burn."

"Burning" the Subject

In both car and truck tails, one of the biggest problems the new agent must overcome is that of mental attitudes. All tail men would readily agree that early in their career they have experienced the feeling that they have "burned" the subject and that the subject is aware of the tail. Experience has shown, however, that this suspicion is usually unfounded and the subject has not in fact been "burned." In case after case, experienced tail men will point out that even though they had become convinced that the subject was aware of the tail, later interrogation brought out the fact that he was never aware of being followed. Only the director of a surveillance operation can keep this phenomenon in balance and be able to temper the agent's mental attitude.

As a rule of thumb, the subject who is actually "burned" will usually resort to illogical or seemingly pointless driving tactics in order to confirm in his own mind the probability of a tail. It is at this point that the agent must drop the tail immediately. The very dropping of the tail will confuse the subject further, as he is unable to confirm his initial suspicion. If the tail is not dropped, then the next logical maneuver on the part of the suspect would be evasive driving tactics which would tend to "lose" the tail.

FOOT TAILS

Unlike vehicular tails, which must constantly consider the "picture" being presented to a rearview mirror, foot tails are more concerned with simply keeping the subject in sight and using good common sense so as not to "burn" the case.

In a foot tail, the agent must be prepared for almost any contingency. He should always have ample change in his pocket for the purchase of a newspaper or magazine and the making of phone calls. Also, he must possess sufficient funds in the event he must utilize public transportation or travel by taxi. Changes of appearance are also desirable on foot tails, probably even more so than in an automobile tail, as the agent is in the open and more exposed. Here again, a cap, a collapsible hat which may be placed in a pocket, reversible top coat, sunglasses or other eyeglasses, are all props which can greatly aid the agent in maintaining his cover. More traditional disguises, such as false beards, mustaches and wigs, generally should not be adopted unless the agent is an expert in their use and the items are of excellent quality so as to present a real-life appearance.

In teaching beginning agents how to conduct foot tails, one of the tactics which can be employed is the traditional "picking a spot" on the rear of the subject and concentrating on that spot when following in crowded conditions. Reflections in storefront windows can also be made use of, but at the same time the agent should be aware that an astute subject who suspects a tail will also utilize the same window angles and reflections to confirm his suspicions.

Another tactic used by suspicious subjects is that of "rounding," which simply means making an abrupt U-turn on the sidewalk and retracing one's steps. The purpose of this, of course, is to attempt to confirm the existence of a tail. In such an event, the agent has little choice but to continue straight ahead, at least until the opportunity presents itself to turn a corner or to enter a store as a shopper would. In any event, during such a "rounding" operation, or on a public conveyance, or at any other time, the agent should *always* avoid eye contact with the subject. For some reason, eye contact more than anything else will tend to confirm the existence of a tail in the subject's mind.

STATIONARY SURVEILLANCE

Hotel Surveillance

Hotel surveillances are probably among the most difficult to maintain successfully. Larger hotels give more cover for the agents, but on the other hand also make it more difficult to locate the subject. Smaller hotels, conversely, offer little or no cover for the agent but do make it easier to spot the subject. If the subject is not known to the agents by sight, then obviously they must work with a photograph. Since the subject's appearance may have changed since the photo was taken, the task may seem almost impossible.

The attitude of most first-class hotels is one of noncooperation in the surveillance of their guests. However, it is often possible for the director of such a surveillance, utilizing personal friendships and contacts, to at least gain the passive cooperation of the hotel security chief. By "passive cooperation" we mean that the agents themselves will not be harrassed by the hotel security staff even though the security staff does nothing to aid the agents in their work.

On rare occasions, it may be possible for agents to rent a room directly across the hall or at either side of the subject's room. In this way, the job becomes much easier. Utilizing radio communication, the agents are then in a position to become aware of the comings and goings of the subject and to be able to pick up on the street surveillance at the hotel entrance. If it is not possible to gain access to a nearby room, the agents can only resort to frequent trips past the subject's room, listening for sounds (or the lack thereof) from within the room, and "plugging" the door. "Plugging" the door means the insertion in the door jamb of a match, toothpick, thread, or other small inconspicuous item which will drop from place when the door is opened.

Use of Vans, Campers, and Buildings

Setting up a surveillance in a business district is usually an easy matter. Even if parking considerations are a problem, the initial point of contact can be handled by one agent on foot with radio communication to the back-up surveillance. In a residential neighborhood, the point of pick-up for the surveillance team can become a problem. Not only are neighbors apt to become alarmed and call the police at the sight of a strange vehicle parked in a residential neighborhood; they may very well become suspicious to the point that loose conversation will eventually find its way to the suspect and alert him to the surveillance.

If an automobile must be utilized in a residential neighborhood, then the surveillance director may decide that he will have to accept the possibility of suspicion on the part of the neighbors. But at least he can attempt to divert the direction of that suspicion. For example, the common tendency is to suspect that the location being watched is in front of the surveillance vehicle, in the direction towards which the agent or agents appear to be facing. In other words, the surveillance director can present this type of a "picture" to neighborhood residents, when in fact the agents are watching a pick-up point to the rear of the surveillance vehicle via the use of sideview and/or rearview mirrors.

The optimum solution to residential surveillance is the use of a surveillance van or camper-type vehicle. After a day or two, such a vehicle parked in a residential neighborhood draws absolutely no attention. Most of the neighbors conclude that it belongs to some visiting friends or relatives of a local resident. The use of well-equipped surveillance vans has gained in popularity in recent years, especially in corporate security departments. One large mail order house with numerous retail outlets across the country is reported to have about fifteen of these vehicles in use. Although vans are generally not desirable for a moving surveillance, especially of automobiles, they seem to be the perfect answer to the need for cover on a stationary surveillance. Campers easily serve the same purpose; the only drawback is that there is generally no direct means of access from the inside of the camper to the cabin of the truck.

In some business districts, where a long-term stationary surveillance is desirable, it is often possible to take an informal rental of a small store building or an unused office. Most rental agents will often agree to the temporary rental of such a unit, especially where the transaction is handled by cash. In this type of situation, the security agent must be alert to the possibility that the rental agency may show the premises for possible legitimate leasing. The surveillance agent's furnishings, therefore, should be kept to an absolute minimum.

The use of a proper cover for a stationary foot surveillance, such as would be maintained during a "buy," is only limited by one's imagination. Here, the cover could easily take the form of a small construction project, street cleaners, taxicab drivers, fishermen at a dockside, sidewalk distributors of leaflets or flyers, or any other suitable arrangement.

Coordination With Police

The question of how far to bring the local police into a private investigation can be a thorny one indeed.

The problem arises because the police, in their efforts to do a proper job of patrolling, are bound to come into contact with stationary surveillance teams or may simply respond to telephone complaints from local residents. Under no circumstances should members of a surveillance team ever reveal to squad car personnel the target of their surveillance. Normally, it should be sufficient for the agents to identify themselves and state the fact that they are on a surveillance but are not at liberty to reveal the person against whom the surveillance is directed. In some small rural communities it may be advisable for the investigator not to show any official identification but to present, instead, his driver's license and attempt to utilize a pretext story which may satisfy the responding police patrolman.

If the security executive decides initially that the local chief or other high-ranking police official can be trusted, then an arrangement could be worked out whereby the surveillance members could simply state to the responding patrolman, "We are on a special project, with which Chief Jones is thoroughly familiar." This, of course, is not completely satisfying to the patrolman, but on the other hand he usually will think twice before challenging that type of statement when it implies that his own chief is privy to the situation.

One of the problems that comes about through the activities of the police patrol is the fact that in stopping the surveillance agents, the police view them as suspicious persons who are loitering in the area and may require them to get out of the vehicle and undergo a search. They may even call for an additional back-up unit with the intention of removing the agents from the area to police headquarters for further questioning. All of this activity tends to attract the attention of neighbors in the community and, of course, that makes it impossible to use that location again.

In a large police department, prior arrangements can sometimes be made through a high-ranking commanding officer so that instructions can be issued to the patrolmen on the beat that a surveillance car with private agents is going to be positioned in the area of a particular block. The block so designated may not be the target of the surveillance itself but an adjoining block; even though a certain amount of suspicion is created on the part of the local residents, it will not be apt to filter back to the suspect.

Experience has shown that when agents are stopped by a police car for speeding, upon displaying proper credentials and explaining to the officer that they are on a tail job involving a larceny, in most cases they will be allowed to proceed without a citation being issued. Here again, the agents should not reveal the type of vehicle which is being tailed, for there is really no need for this; if the officer presses for this information, it is only because of his personal curiosity.

If the surveillance agents are on a stationary stake-out of a particular location and they are approached at night by a police officer, their first act should be to turn on the interior lights of the car. Most policemen are extremely wary of approaching one or more occupants in a parked car at night and usually do so only when they have unholstered their weapons or at least removed the safety snap from the holster. The sudden appearance of an interior light in the automobile tends to reduce the anxiety of the approaching officer and convey the message that the agents are legitimate and that everything is open and aboveboard.

Communications and Special Equipment

Coordinating with either inside or street undercover agents, the tail men should work out in advance a contact point whereby messages can be relayed via telephone. Often the undercover agent may become aware of last-minute changes in the suspects' planning, and unless a prearranged line of communication is set up, he will be unable to convey this information to the surveillance agents. The undercover agent may also be necessary to identify or to "finger" certain suspects in the case. This, of course, has to be worked out in advance with the surveillance personnel so that they may "pick up" the suspect. (As used in surveillance work, the term "pick up" is not to be confused with its use as a synonym for apprehension. It simply denotes the beginning of a tail job of a particular subject at a specific location.)

Prior to the advent of modern-day radio equipment, hand signals were often the only way to accomplish the leapfrog concept of multiple-vehicle surveillance. Today, nationwide companies are able to obtain a designated frequency for the use of UHF radio equipment. A good radio communication consultant is really the key to the solution of radio problems and their legality anywhere in the country. One major corporation has been able to secure a radio license covering every major location point within its industrial complex in the United States. This enables its corporate security department to move from locale to locale and still use its UHF radio equipment legally.

Earlier, reference was made to the use of a specially equipped surveillance van. A number of security departments have purchased stock vans from major auto manufacturers and modified them for their own use. If such a van is to be disguised in some way, then proper consideration should be given to the van's "cover." For example, the van may have signs describing it as belonging to an industrial testing company, an engineering survey company, a pollution study firm, etc. If such cover is utilized and signs are affixed to the van, then the cover should be complete with telephone numbers that at least answer and identify themselves as an answering service.

The interior of the van should be pretty much tailor-made to

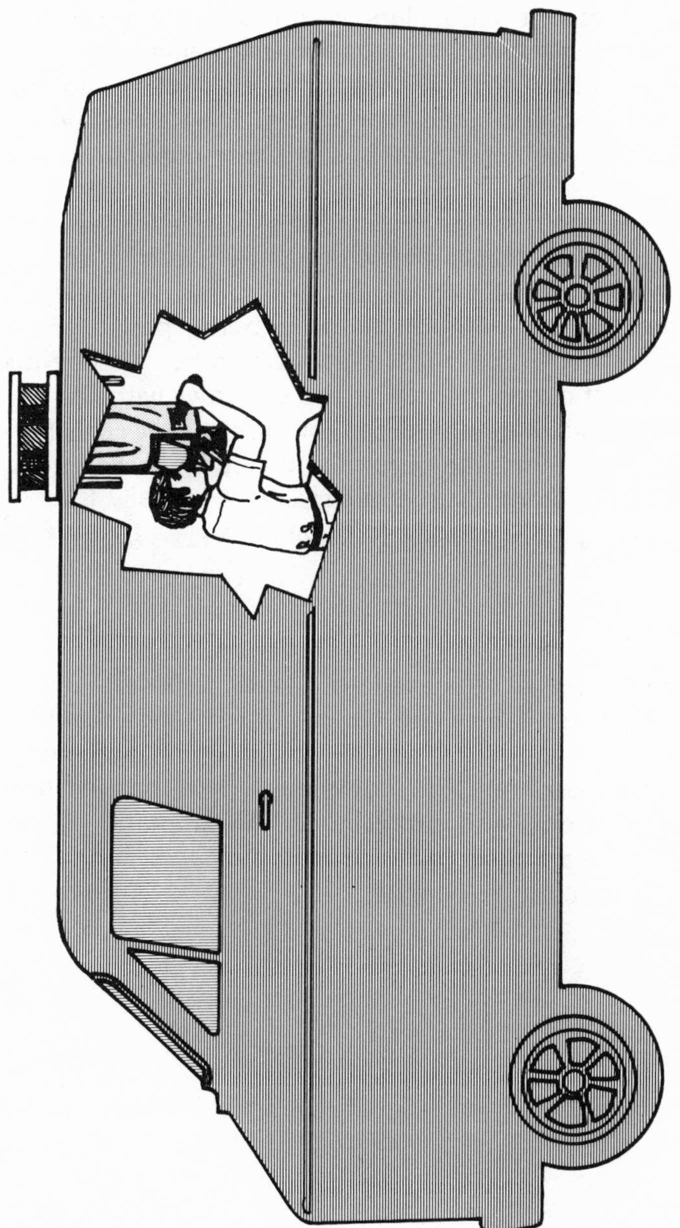

Figure 6-1. Disguised in an air vent, periscope enables occupants of surveillance van to observe subjects without detection.

the needs of the operation but at least should contain proper camera equipment, binoculars, chemical toilet, and food and beverage storage which would enable the agent to remain in the van for at least several days. Obviously, this would predicate the equipping of the van with at least a folding cot, a small table for writing reports, and other emergency equipment.

An innovation that has been made available for surveillance vans is a type of periscope which enables the occupants of the van not only to observe their target but also to make use of their photographic equipment through the periscope. (See Figure 6-1.) Lacking a periscope, the van can also be rigged with black-out curtains on all of the windows with the exception of the window being utilized for observation. This reverse lighting effect enables the occupant to view outward through a curtain of linen composition but makes it virtually impossible for someone on the outside to be able to view the interior of the van.

Conclusion

Obviously, not every case lends itself to successful conclusion through the use of the various surveillance techniques. On the other hand, there are always a certain percentage of cases which can only be completed in a proper fashion through stake-outs and surveillance. A command of the intricacies of the art of surveillance is an indication of the overall ability of the security executive.

Chapter 7

Supplementary Investigative Methods

Many cases are of such a nature that neither undercover work nor various forms of surveillance are appropriate investigative approaches. In other cases these approaches fail to produce the necessary information to conclude the matter. This would be especially true of high-level executive crime, where undercover investigation would be totally out of the question and various surveillance attempts would probably produce only limited results. In these types of situations, the investigation manager must turn to other avenues of information to develop his case.

Agency Background Checks

In the case of investigative subjects with interstate backgrounds, a good starting point might be a background investigation conducted by a national agency specializing in this type of work. A number of good agencies perform this service. Agencies of this type are able to pool various investigative requests into certain geographical areas, resulting in a savings to the client, unless a priority is requested so that all investigative effort is directed to a single request. These agencies have investigators who regularly spend parts of each working week in courthouses, in close contact with various credit bureaus, and also have occasional regular contact with educational institutions and the like. The use of such an investigation agency is a great time- and money-saver to the investigations manager. It will provide at least an overview of the subject's background and may very well indicate leads of questionable areas that should be pursued in a follow-up investigation.

Most background investigations of the type mentioned above cost a minimum of $250 and may cost as much as one to two thousand dollars, depending on the information requested and the number of geographical points and contacts that must be made. The important thing to be remembered in using this approach is that most of the information is obtained through direct personal contact. There is little use of mail or telephone, as there is by some agencies offering background investigations in the neighborhood of twenty-five to fifty dollars. Anyone experienced in background investigations will readily appreciate the advantage of spending the extra money to obtain the personal interview of past employers and credit sources, rather than using the faceless approach of the mails and telephone.

In-House Supplemental Investigation

After receiving the results of a nationwide background investigation, such as described above, the security manager is often able to zero in on loose ends which need to be followed up or even leads or questionable areas which may have been developed by the original investigative agency. These follow-up investigations are often more productive when performed by the security manager or his staff. This is especially true when it comes to a high-level interview which may be conducted with a top-ranking executive in another corporation. Experience has shown that often a corporate executive will be inclined to be more candid than he was with the representative of the security agency which conducted the initial check. Corporate executives may be less than candid when talking to an agency investigator because of fear of libel or slander suits.

In summary, the background investigative agency can usually act as a tremendous time- and money-saver to the security manager by simultaneously working investigative or background leads in various parts of the country. This approach enables the security manager to make an intelligent decision regarding the area or locale on which he will focus his own in-house effort.

Bribery and Kickback Cases

To most corporate security managers, the most difficult and frustrating cases to bring to a successful conclusion are bribery and kickback cases. In many industries, bribery or kickbacks are a way of life, something that is taken for granted by executives in that particular industry.

Company Policy. Some companies attempt to dampen the kickback problem through the use of periodic written communications to their suppliers. These communications generally state the company's policy as to honest business practices, suggest tactfully to the supplier that kickbacks are unnecessary to do business, and advise the supplier that if kickbacks are solicited by a particular executive, they should be reported. Coincidentally with this, many of these same companies will make use of a conflict of interest questionnaire which must be completed and signed annually by all key executive personnel.

Such deterrents may serve well in the case of a basically honest executive who might otherwise be tempted. On the other hand, they are of little value to the corporation with a dishonest executive who has been solicited by a dishonest supplier to engage in a kickback arrangement.

Difficulty of Detection. There are statutes at the state and federal levels which cover bribery and kickback situations involving public servants in their relationships with outside contractors or service companies. Most of these laws contain serious penalties for conviction and are on a felony level. For executives of private companies, however, it is generally not illegal to engage in such behavior. Only a few states have seen fit to pass commercial bribery statutes. However, virtually all of these are classified as misdemeanors and in reality would probably draw absolutely no jail time whatsoever.

The truth of the matter is that unless either the "bribee" or the "briber" decides to reveal the existence of such a practice, the kickback scheme may very well go undetected; even if suspected, it may be impossible to prove it. Some of these kickback cases have been brought to light by a close business associate who was not involved but who knew of the scheme. However, in virtually all of the situations, the company executive who was the kickback recipient was a long-term employee with much to lose by admitting to the scheme. Unless there is a falling-out between the two participants and one comes forward and makes such an admission, the only other approach that can be used is interrogation. Interrogation should be undertaken only after as many facts as possible have been developed through ordinary investigative methods.

Outside Agency Investigation. Here again, as with background investigations, a personal investigation of the suspected recipient can be conducted by an outside agency. This is generally preferable to an in-house approach, because in any investigation of this type there is always a certain risk that the subject of the investigation will learn of the investigative efforts. By employing an outside agency, the security manager at least sets up a layer of insulation between himself and his co-executive who is being investigated.

Even when utilizing an outside agency, the security manager should exercise close control over the investigation and make the decisions as to what investigative sources will be checked and what persons will be interviewed by the agency. A surveillance of the suspect may prove beneficial, along with a covert inspection of his business quarters after hours. In some cases, the security manager may be able to use personal contacts which he has developed with his counterparts in various stores and utility agencies to obtain information relating to current accounts, long distance telephone calls, and other useful information.

Internal Auditing Department. Another useful approach in attempting to develop investigative leads is through the use of the internal auditing department. It is generally expected by most executives that internal auditors perform periodic routine examinations of certain business records and financial information. To make use of this ploy for a specific examination of a suspect's company accounts is an excellent pretext. The auditor is often able to come up with additional leads through an examination of the suspect's expense reports, travel records, etc. Even if the security manager wishes to make the examinations himself, the auditor is in a much better position than a security representative to secure such documents without arousing suspicion.

Recipient's Own Records. In kickback cases the recipient almost always keeps an independent record of the goods purchased and the margin of kickback which is applied. This may take the form of a small diary kept in his desk, or loose sheets of paper in a folder, or a three-ring loose-leaf notebook kept in the trunk of the recipient's car. Through the use of his own resourcefulness, ingenuity and imagination, the security manager will occasionally be able to conclude a kickback case successfully.

Interviews With Witnesses

As the typical investigative case nears its culmination, the security manager may be tempted to resort to interviews with witnesses in order to strengthen the facts already obtained. As one can surmise, overt interviews with witnesses are usually more productive than interviews by pretext. By the same token, they are also more risky than pretext interviews. In most companies today, there is still a small cadre of employees who possess a loyalty to the company which transcends any loyalty to their fellow workers. It is these people who are often more susceptible to overt interviews than any others. The basic company loyalty of such persons is only enhanced by an open, direct interview with a top security representative. It should be kept in mind that this is an experience that most employees never encounter, and many witnesses actually feel a sense of flattery that the security department of a large corporation would seek them out and spend time with them as witnesses during an interview.

With other types of employees or with outsiders, an interview by pretext is often the only safe approach. Security representatives must be familiar with the laws of the state in which they are operating to insure that there is no prohibition against interviews by pretext. Some common pretext approaches are insurance risk investigations, job applicant or background investigations, consumer and demographic surveys, and poll-taking. Here again, the approach to be used is only limited by one's imagination and whatever legal restrictions apply.

The timing of such interviews, whether overt or by pretext, is all-important. There are risks involved in both approaches, and the security manager must weigh these risks against what there is to be gained. The risks are often minimized by a very close tie-in between the interview and the actual breaking of the case itself. In these instances the interview may very well be worthwhile. The following case example illustrates this point:

New York, New York . . . . Based on information received from an informant, the security manager of a large national company realized the existence of a kickback scheme in the sales department. Basically, sales trainees were required to pay "tribute" to their supervisor, who, after taking his cut, passed the remainder on to the regional sales manager and to the national sales manager. The money was generated by falsifying travel expense reports under the guidance of the territory salesmen. The trainees were allowed to keep half of any excess funds, so that they, too, quickly became compromised. For years this particular sales region had been used as a training ground for the rest of the country.

It was decided to interview initially all former trainees who had left the company, just prior to interrogating the two top executives. In order to maintain the element of surprise, key security representatives left New York on a Friday evening, fanning out across the country. Over Saturday and Sunday, former trainees were interviewed in such diverse places as Atlanta, Oklahoma City, Dallas, Los Angeles, Chicago and many other cities. Signed statements were obtained from virtually all of them and the interviewers were back in New York by Monday morning. On that day in question, both executives were intercepted for interrogation as they arrived for work. Later events confirmed that only one executive had been forewarned, and this only by a single phone call that was very general in nature and had not revealed any specific admissions.

The value of a thorough interview should never be discounted. It should be kept in mind that among the most successful skip tracers in the country today, good interviews constitute probably fifty per cent of the successful techniques utilized. (A skip tracer is an investigator who specializes in locating persons who have "skipped"—that is, absconded to avoid payment of debts or other obligations.)

Covert Facility Inspections

Unlike overt inspections of company facilities which are intended to act as a deterrent to business crime, the covert facility inspection is designed to develop evidence without alerting the suspects. Common sense dictates that such inspections of business facilities should always be conducted after hours when the likelihood of discovery is at an absolute minimum. Here again, the investigator is looking for corroborating evidence which could be found in personal records, diaries, desk calendars, etc. In the case of non-executive personnel in a factory or distribution center facility, the investigator could easily be looking for contraband in lockers, caches of merchandise in work stations or out-of-the-way locations, narcotics or dangerous drugs, gambling paraphernalia, or anything else that would assist an interrogator in overcoming the resistance of a suspect and obtaining a written confession.

SECTION III

THE "BUST": CLOSING THE INVESTIGATION

Chapter 8

Preparations for the "Bust"

Having discussed in Section I the various indicators of theft in the business or industrial setting, and in Section II the investigative methods used to further pursue these indicators, we arrive at a crossroads faced by many companies when the probability of extensive internal theft suddenly becomes a certainty and a reality. A decision must be made by top management as to whether entrenched thieves should be rooted out; whether only one or two should be singled out as an example; or whether added controls and restraints should be implemented within the company to make it more difficult for the thieving to continue.

Many veteran security chiefs have come to the conclusion that added controls and restraints are intended for workers who are basically honest but who might otherwise be tempted by loose procedures and an immoral plant atmosphere. They further believe that employees who have developed such patterns of theft as to be classified as professional or semi-professional thieves cannot be easily rehabilitated or even deterred by added controls. In these cases the only solution is to identify thieves, separate them from the company payroll and, where possible, prosecute.

Experience has shown that many corporate security managers and directors of investigative agencies are extremely reluctant to tackle large theft groups in a company facility. Many security heads have not had the experience or even the opportunity to close up a case involving large numbers of confirmed thieves. Accordingly, Section III of this text will emphasize the techniques necessary in handling a large organized theft group, as opposed to more conventional methods that might be utilized with one or two employee thieves. The successful techniques which will be discussed have evolved over a period of time and are both proven and flexible, capable of meeting changing and different conditions. The plan of action to be developed in the following chapters will work each and every time, if the security chief has the authority, intestinal fortitude, and ability to carry out the detailed planning which must necessarily be part of the case.

Admittedly, some of these concepts can be classified as a radical approach, but it is one which can be likened to radical surgery for cancer. Many authorities have come to view widespread theft in industry as an industrial cancer, and in a union shop, the "surgeon" or security chief must realize that he probably will get one shot at the disease and one shot only. In the author's experience, the concepts which follow have proved successful in 51 out of 52 attempts—successful in that signed statements or other evidence sufficient to sustain discharge were obtained from a majority of the suspects. In the only failure experienced, the scene of action was outside the United States and the closing of the investigation was made impossible because of strong nationalistic feelings on the part of the perpetrators.

Picking the Date and Time

The planning involved in an operation which seeks to close a case on a group of ten to twenty or more hard-core thieving employees can be likened to the planning required for a military operation. Where large numbers of suspects are involved, the "bust" will always have some effect on the day-to-day operations of the company, and for this reason top management must be in agreement as to the least disruptive date from a production standpoint. In order to permit peak shipping schedules to be met, for example, the date chosen might be as much as thirty days beyond the closing investigatory days of the case.

From the security chief's standpoint, the day of the week on which the closing operation begins is extremely important. The case must move rapidly and be concluded before the end of the normal work week. Therefore, a Monday is highly desirable; certainly a closing should be commenced no later in the week than Tuesday if large numbers of people must be dealt with. The time of day is also extremely important. The investigators must be able to utilize as much of the normal work day as possible for their closing operation, as it is often difficult to operate on an overtime basis beyond the plant's normal quitting time.

Need for Detailed Planning

In planning for the closing of a large-scale internal theft investigation, the security chief should view the entire operation as a military type exercise, with the enemy represented by the hard-core thieves or theft ring. In a militant union situation, the union must also be recognized as a potentially dangerous ally (either willingly or unwillingly) of the theft group.

Many union locals have long viewed company security departments as "tools of the bosses" and have given security personnel such epithets as "Company Gestapo" or "Company Pigs." Accordingly, during the course of a major investigation or attempts at interrogation of rank-and-file employees, many union locals will immediately jump to the aid of the employees by advising them of their "rights." In the past, some union locals have issued statements to the workers advising them to refuse to talk to security personnel, to refrain from making any admissions or signing any statements, or even to refuse to be interviewed without the shop steward being present. A number of unions have incorporated into their constitutions or bylaws prohibitions against one worker implicating a fellow union member. All of these factors work to the detriment of the security chief and greatly impede an investigation. Officially, virtually all unions take a stance against employee thievery. Realistically, the most that a security director can hope for is that the union local will remain passive throughout the investigation and enable him to complete his task.

With all of the above considerations in mind, it becomes clear that in order for the closing of the investigation to be successful, the planning must be detailed and sufficient in scope to enable the security personnel to excise the cancer before the entrenched thieves or the union have an opportunity to raise defenses to the operation.

Investigation Summary

After picking a time and date for the closing of the investigation, the security director must now summarize and cross-reference all of the information which has been developed throughout the investigation. Normally, the bulk of this data will be extracted from undercover reports, but the case file may also very well include the results of successful surveillances, tail jobs, or other investigative efforts.

Most undercover and surveillance reports are written in a chronological order based on date and time. In the investigation summary, pertinent information is extracted according to each suspect and then reassembled in chronological order under each suspect's name. The undercover information can easily be integrated with outside surveillance reports and information from any other sources.

The purpose of the investigation summary is to provide each member of the interrogation team with a synopsis of all pertinent information which has been developed on each suspect in the case. As the interrogator ultimately is assigned various suspects for interrogation, he will immediately have at his fingertips every bit of information known about each particular suspect. This gives an interrogator a tremendous advantage and easily accounts for the confession success rates of 90 and 95 per cent which are common in investigation closings of this type. Psychologically, the appearance of an investigation summary, which has been typed and may be contained in a folder, is devastating to the suspect. With this type of assistance, a skilled interrogator can completely destroy the psychological defenses which normally are raised by a suspect in this kind of situation.

In compiling the investigation summary, the security director must be careful not to expose the existence of his undercover agent. Normally, the summary can be written in such a way as to make it appear that conversations held with the undercover agent by various suspects were overheard by a third party. In addition, employees can often be led into believing in the existence of hidden microphones and cameras.

Realizing that the various interrogators may not appreciate the delicateness of the undercover investigation, the security director must make every effort in writing the summary to conceal the true identity of his undercover investigator. Furthermore, it must also be realized that the investigation summary may ultimately fall into the wrong hands, either inadvertently or through legal process; all the more reason that the identity of the undercover agent must not be compromised. The following are examples of investigation summaries from actual cases.

B____, Ed

Employed as order filler, 1 year. Subject drives a 1965 white Cadillac convertible.

12-1-70 Subject overheard stating to other employees that he had been stealing shirts and placing them in his automobile.

12-14-70 Subject overheard telling another employee that most of the shirts which he stole were taken out of the building on Saturdays.

1-21-71 Subject was overheard stating to another employee that he had agreed to steal 30 shirts for L____ W____ on this date. However, L____ did not come to work.

1-22-71 Subject was overheard stating to another employee that he had a part-time job working at night at the ______ department store in Marietta. He was overheard to state that he had stolen about $500 worth of goods since employed there.

1-27-71 Subject observed stealing 2 shirts and placing them in his car during lunch period. These were placed under the floor mat, as per usual. At quitting time, subject observed stealing 4 additional shirts. It is to be noted that subject seems to steal most of the shirts when he wears a blue jacket.

2-2-71 Subject overheard discussing with another employee the possibility of slipping 3 or 4 additional cases of shirts on the truck which makes a shuttle run between warehouses. The gist of the conversation was that the seal would not be put on tight and that the shirts could be dropped at a location on McAfee Street.

2-4-71 Subject and another employee were observed slipping an extra case of shirts on the shuttle truck (3-8/12 doz.). Outside surveillance developed the fact that shirts were dropped off at Apt. 620, 5 McAfee Street, Atlanta. After the shirts were dropped off it was noted that B____ actually locked the seal on the truck. At quitting time on this date subject was observed stealing 2 additional shirts from the warehouse and then selling them to employee L____ D____ outside the warehouse.

2-15-71 Information received that subject may, on occasion, carry a pistol which he has stolen from the ______ store. Subject overheard in conversation this date stating that he had stolen a total of 3 pistols.

Outside surveillance indicated that several extra cases were loaded on the shuttle truck to which subject was assigned at the M____ warehouse and that these extra cases of shirts were dropped off at the address at McAfee Street. It was noted that subject and another employee returned to the McAfee address after quitting time and apparently made a split of the loot. Subject was also overheard during the day at the warehouse to state that he assisted another employee who worked in the Military section to remove 5 shirts from the warehouse.

H____, Willie Frank

Terminated approximately 6 weeks ago for tardiness. Drives a '69 or '70 sky blue Volkswagen.

10-23-70 At approximately 1:00 p.m. subject was observed examining the contents of different shirt boxes and at one point was observed placing a shirt down inside his trousers. After leaving the area he returned and repeated the procedure, taking another shirt.

10-26-70 Subject observed placing 2 shirts in his trousers during the afternoon. It is estimated that at the rate of theft which has been observed that subject may take as many as 20 to 30 shirts during the course of a particular day.

10-27-70 Subject observed again during the afternoon on this date stealing a large quantity of shirts.

10-29-70 Subject overheard expressing concern about the removal of a cache of 50 shirts which he had in the warehouse. This resulted from a security meeting held by management for the employees.

11-13-70 Subject was seen showing a stolen shirt to an-

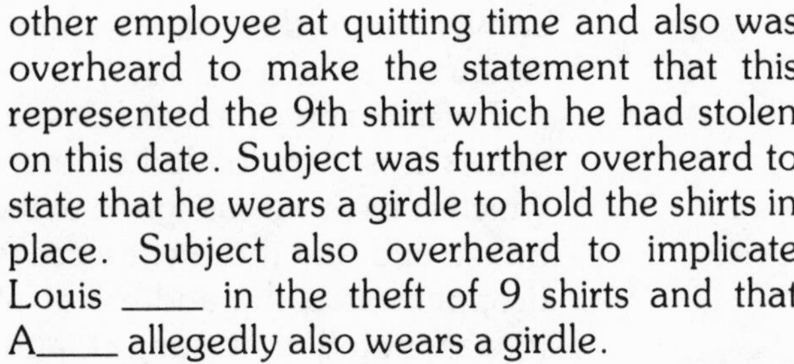

other employee at quitting time and also was overheard to make the statement that this represented the 9th shirt which he had stolen on this date. Subject was further overheard to state that he wears a girdle to hold the shirts in place. Subject also overheard to implicate Louis ____ in the theft of 9 shirts and that A____ allegedly also wears a girdle.

11-24-70 Subject observed in the company of Willie ____, Tony ____ and Johnny ____ stealing a large quantity of shirts and placing them in the pockets of their coats.

11-30-70 At approximately 4:00 p.m. subject observed placing a shirt inside of his pants. This was witnessed by a picker named Paul ____.

12-16-70 Subject observed removing 2 shirts from under his trousers and placing them on the back seat of his car at quitting time. This was witnessed by employee Leroy ____.

12-17-70 Information developed this date which would indicate that subject is working in collusion with employee Leroy ____. Subject was observed removing 4 shirts from his pants and placing same on the back seat of his car. Later, subject stopped at a "drop," which was an old wrecked car, and took out 2 more shirts. These had apparently been placed in the wreck during the noon hours.

12-18-70 Subject overheard stating to another employee that he had only stolen 4 shirts on this date.

1-4-71 Subject observed taking 1 shirt on this date and was overheard telling another employee that he had an order to get 25 more shirts by the following Friday for $125. Subject stated that he had anticipated taking most of them during the lunch hour.

1-5-71 Subject observed stealing 4 shirts on this date.

1-6-71 Subject observed stealing 4 shirts on this date. Subject was overheard boasting that he was becoming so proficient in theft that he was able to steal shirts without taking off the wrappers.

1-7-71 Subject observed stealing 2 shirts on this date.

1-8-71 Subject observed stealing 3 shirts at quitting this date.

1-11-71 Subject observed in theft of only 1 shirt this date. Subject was overheard telling another employee that he had obtained $130 from the deal which he made the previous Saturday. He further stated that his contact wanted an additional 25 shirts.

1-12-71 Subject observed in the theft of 2 shirts this date.

H____, Tom

Subject drives a dark blue 1970 Volkswagen which is usually parked near the back fence. License #BB-183.

10-20-71 Subject overheard announcing at lunch time to the other employees that he had "speed" available for sale.

11-6-70 Subject believed to have supplied Kenneth ____ with a quantity of "hash" on this date.

11-12-70 Subject overheard offering to sell either opium or "speed" to another employee.

11-13-70 At approximately 2:45 p.m. subject was observed selling an unknown quantity of "hash" to another employee for $8. This transaction took place in the men's restroom.

12-16-70 Subject overheard in the men's restroom offering to sell 70 hits of "speed."

1-11-71 Subject, along with two other employees, observed in the men's room smoking marijuana. Subject is now believed to be in business with employee Ernest ____ in the selling of drugs and narcotics within the warehouse.

1-29-71 Subject observed at about quitting time selling 3 capsules of "speed" for $3.

2-5-71 Subject observed selling some "hash" to another employee. Also overheard stating that he was going to attempt to get supervisor Skip S____ to buy some illicit drugs.

3-3-71 Subject observed selling "joints" for $1.50 a "stick."

O____, Clay

10-28-70 Employee Jim ____ overheard in conversation that another employee implicated subject in the theft of 3 new large ashtrays the previous night. It was alleged that subject had them under his coat when he went out at quitting time.

10-29-70 Subject involved in steady drinking on the 3:00 to 11:00 p.m. shift on this date in the company of Jim ____, Roger ____ and others. Subject brags of multiple arrests for drunk and disorderly. After the drinking episode the men re-entered the plant at approximately 10:30 p.m. and subject made a notation on the Tenting machine pad, "Down 9:45 to 11:00 p.m.—Machine overheated." Then subject was observed turning all of the dials on to high, which in fact caused the machine to overheat. Subject also observed punching out Roger ____'s time card this date.

11-7-70 Subject engaged in heavy drinking on this date. (See entry on William ____). While intoxicated, subject stated that he had murdered his Sergeant while in Vietnam by shooting him in the back of the head while on patrol. He also admitted that he had had apparent psychiatric problems in connection with his military experience.

11-12-70 Subject overheard in conversation with other employees admitting that he had taken home a whole box of masking tape.

11-17-70 Subject was overheard this date telling another employee how he causes false readings on the Totalizer by hitting it and causing the numbers to jump as much as 1,000 yards each time.

11-25-70 Subject believed to have been picked up by the Saratoga police at the plant on this date. (To be verified later.)

11-30-70 Subject participated in a theft and drinking orgy on this date along with others. He was directly involved in theft of a can of Babo, allegedly intended for William ____, a tape measure, and 2 slips from Mr. H____'s office.

12-4-70 Subject made the statement that William ____ needed some material #AN402 for his apartment for drapes, bedspread and blankets, as well as to cover furniture. Consequently, several of the men took off 14 to 15 yards of material, cut it in half, wrapped it, and laid it aside. Then subject obtained 6 or 7 yards of the material for his own use. Later, subject took approximately 8 to 9 yards of material #290 and obtained help from another employee in wrapping same. Later in the evening, subject attempted to retrieve the original stolen material intended for William ____ but discovered that half of it had already been stolen and expressed the thought that it had been taken out by ____. The material that remained was taken out by subject, past the guard, and unwrapped. The guard said nothing.

12-7-70 At approximately 10:55 p.m. subject was observed punching out the time card of Roger ____, who had left the plant at 6:00 p.m. in an intoxicated condition.

12-9-70 Subject was observed at quitting time carrying out 5 to 6 yards of material #402 which had been obtained earlier off the end of a run.

12-10-70 Subject was observed punching out the time card of an unidentified employee at quitting time. It is believed that the other employee had left work at approximately 8:15 p.m.

12-14-70 At approximately 10 p.m. subject was observed in the company of Tom ____, while he obtained approximately 12 to 14 yards of a patterned Monofil material, approximately 50 inches wide, from the scrap box. Subject wrapped same in brown paper.

12-15-70 Subject was seen punching out the time card of Jim ____ at 11 p.m. Jim ____ had left the plant at 9 p.m. but had had an automobile accident at 9:40 p.m., which was subsequently printed in the local newspaper.

12-17-70 Subject obtained liquor and other items for a party on this date with $18 which was given to him by Floyd ____. After becoming intoxicated, subject was overheard to admit to another employee that he had stolen an unknown quantity of masking tape for a painter in Saratoga. Subject is believed to be responsible for breaking the lock on the door of the office in which the party was held on that night.

Manpower

One of the early problems that must be given consideration by the security director is that of manpower for the closing of the investigation. The number of security personnel assigned will depend on the number of suspects and whether or not the company has been organized by a labor union. The short-range objective should be to get through with preliminary statements from the bulk of the prime suspects by the end of the first evening. If this can be accomplished, then the security director has sufficient psychological momentum going into the second day so that union efforts to thwart the investigation should prove largely ineffective. For example, if the investigation summary indicates a total of thirty possible suspects, twenty of whom can be considered prime suspects, then the investigation team should contain at least eight or ten personnel.

Most of the members of the team should be expert interrogators, able to perform in a low-key manner, in a one-on-one situation. At least one member of the team should be a qualified polygraph examiner, and another member of the team should be adept at making searches for the possible recovery of contraband. A good secretary or stenographer is also indispensable.

The Scene of Operations

Inasmuch as the success of the operation will depend on complete secrecy and surprise of attack, the scene of operations must not be the plant itself but rather some other location so that the investigation team will not be operating in the suspects' "back yard." By having the scene of operations away from the plant, the chances of maintaining complete secrecy throughout the first day (and thus the chances of concluding the operation successfully) are greatly enhanced.

A good quality motel within a very short driving distance from the plant has been found to be the best solution. The security director must not attempt to skimp on room costs, just as he should not hesitate to bring adequate manpower for the job. Ideally, a large suite should be designated as a waiting room for suspects, with the adjoining bedroom serving as a control center and office for the director. Each of the interrogators should be assigned a room for interrogation purposes, and one room can be set up for polygraph examinations. The motel or hotel should be one which also offers room service. It is desirable to feed the suspects at mealtime in the waiting room rather than attempting to take them into the coffee shop. It is not necessary for the various interrogation rooms to be connecting, but they should be in close proximity to the suite.

In making reservations for such motel accommodations, the security chief must be careful to use a cover story which will preserve the integrity of his plan. An adequate story has always been that of a personnel consulting firm which intends to do extensive interviews of people over a period of two or three days. This type of a cover is logical in nature and is entirely in keeping with the type of accommodations which will be required.

Equipment Needs

The advance planning in an operation of this type also includes a list of needed equipment. If it is anticipated that a fair amount of stolen merchandise will be recovered, then possibly two vehicles should be rented, one of which should be a station wagon which can be utilized to transport bulkier contraband. In some extreme cases where ordinary vehicles would not suffice, it is possible to utilize a company truck on a short-term basis.

Typewriters and other office supplies, including shorthand books, pencils and erasers, will be needed for the taking of statements. In order to process recovered evidence, boxes, bags, labels, staplers and tags should be provided. Kraft grocery-type bags are usually sufficient and can be adapted for use with an evidence stamp which can be affixed beforehand. (See Figure 8-1). Clipboards should be provided in the same number as the number of anticipated suspects. If the undercover agent was successful in marking any contraband with ultraviolet crayons, then of course equipment should include an ultraviolet light to be used in scanning recovered property.

Legal and Union Review

Part of the thorough planning which is required for an operation of this type must include a review of any union

EVIDENCE

BAG NO.______OF______BAGS

SUSPECT:________________________________

ADDRESS:________________________________

PHONE # (   )__________SSN______________

OWNER / VICTIM:__________________________

LOCATION:_______________________________

REMARKS:_______________________________

____

____

I HAVE INITIALED AND DATED EACH ARTICLE, SEALED THIS BAG AND INITIALED THE SEAL.

____________________   ____________________

DATE   **SECURITY OFFICER**

Figure 8-1. Rubber stamp to be imprinted on evidence bags.

contract as well as a familiarization with the local state laws governing larceny, frauds, and/or embezzlements.

Many union contracts have unusual procedures included in sections relating to disciplinary action, grievances and arbitrations. The security chief must be aware of any restrictions in the labor contract which might have some bearing on his investigative closing. As an example, many security heads would be inclined to suspend a particular employee after interrogation and statement-taking, pending a final disposition by company management. Some union contracts, however, do not provide for suspension, and any suspension is thus considered a direct disciplinary action. The security director should avoid, at all costs, participating in disciplinary actions. In cases of this type, the lack of a suspension clause can be circumvented by simply telling the employee to remain at home, on company time, pending final disposition of his case. To remain at home on company pay for a day or two while the investigation is being completed would not be considered a disciplinary action and would not become the subject of a grievance.

The mere interrogation or processing of a union member is not considered a disciplinary action per se and therefore union representation does not have to be offered.[10] At the point of discharge, however, the action does become disciplinary in nature and the union is entitled to (and should) be made a party to the discharge, whether or not the offending employee requests representation.

Many states have a number of variations in the criminal statutes covering larceny and similar offenses. A few states even have special civil rights guarantees to offenders who ultimately are prosecuted. In the state of Minnesota, for instance, an offender must be given a copy of his signed confession at the time it is signed. This holds true even though the confession is obtained by a private or corporate security representative and without any police involvement. Several other states have gone beyond the federal requirements in this regard and have similar provisions.

Although the primary objective of the security director is to separate the hard-core thieves from the company payroll, he should not overlook the possibility of successful prosecution. In fact, if the case is built along the lines of possible prosecution, separation from the payroll becomes much easier and the case has then been prepared in the proper manner in the event of a union arbitration.

---

[10]For a more complete discussion of union representation, see the following chapter.

With a good understanding of the various forms of larceny and embezzlement, the security chief can proceed to build his cases during the closing with an eye toward proper prosecution. For instance, in some states theft from a warehouse is considered an automatic felony regardless of the amount involved. If he is familiar with such a law, the security director can easily word any signed statement to clearly reflect that the thefts took place from a warehouse rather than simply from a company, if this is the case. In other states, theft from a warehouse has been known to be considered the same as theft from a mercantile establishment and thus falls under the larceny statutes covering shoplifting!

If the security director is fully versed in the intricate twists and turns of the state criminal statutes and the union contract, if any, he should be able to build a strong case which will have little chance of being knocked down at a later date because of some technicality.

The Briefing

Assuming that Monday morning has been chosen for the beginning of the "bust," the security manager will undoubtedly wish to move his security crew into the community and the motel on Sunday afternoon. Inasmuch as some of the members of the crew may be outsiders, not directly employed by the security department, it is highly advisable to hold a briefing on Sunday evening so that procedures can be made clear to all security personnel and any last-minute intelligence can be added to the investigative summary.

Prime Suspects. The assembled security personnel should be briefed on the principal suspects, whom they will normally encounter in the first group of employees to be interrogated. Good tactical planning requires that the suspects with the most evidence against them should be included in the "first wave." In this way, the chances are increased of picking up additional information and additional implications of new suspects not already known.

Group Leaders. In addition to prime suspects, the "first

wave" should also include any strong leader of the group, such as a shop steward, who would otherwise be likely to agitate and create problems back at the plant after it was discovered that an investigation was in process. By including the leader in the first wave, the security team has left the rest of the group without leadership and has also severed primary communication lines to any outside source of help for the thieves. The effectiveness of these tactics is exemplified in the following actual case histories.

Oakland, California . . . . Investigation by two company undercover agents of a general merchandise warehouse firm employing approximately 150 workers had continued for approximately six months. At the end of this time, a number of prime suspects had been identified as being involved in systematic theft.

The whole work force was dominated by the single shop steward who constantly attempted to intimidate management and who was frequently successful in his efforts. During the six-month investigation, the only piece of information developed on the shop steward was the fact that on one day he had been seen entering a restricted area of the warehouse where he had no logical reason to be and in so doing had violated company rules. Although there was no other piece of information at hand regarding this shop steward, it was decided that he would be included in the first wave to be questioned at the motel and would be interrogated by the most skilled interrogator on the security team.

The interrogator was given instructions simply to keep a conversation going with the shop steward throughout the day until quitting time. It was hoped that this would break the theft group's line of communication to the union and would enable the security personnel to get the "jump" on the case during that first day's activities.

By noon of the first day, the investigation team had identified two separate theft rings within the facility, both of which were working independently but with full knowledge of the other's activity. The shop steward who had been simply targeted for neutralization turned out to be the head of the larger of the two organized theft rings!

A second case history illustrates the same basic strategy:

Detroit, Michigan . . . . After a number of months of undercover investigation in a drug warehouse, the investigator was able to identify the shop steward in the systematic theft of prescription and other drug items on a daily basis. This shop steward customarily performed only a minimum amount of work and spent most of his time agitating the other employees and keeping the management group in a constant uproar. Over a period of years, he had become both physically and mentally intimidating, to the point where first and second line supervision gave him a wide berth. It was the union steward's practice to leave the plant approximately fifteen minutes earlier than the normal quitting time.

A decision was made to apprehend the steward at the front of the plant when he was in possession of a small quantity of stolen drug merchandise. This apprehension took place late Monday afternoon. Within several hours, written statements had been obtained, the shop steward's home had been searched and additional drugs recovered, the subject was charged and had been incarcerated. The rest of the closing of the investigation was begun the following morning. With no leadership to back them, the members of the theft ring which was operating in the plant were easily handled and processed according to plan.

Assignments. During the briefing, the director should give the other members of the security team their primary assignments. Normally, most members of the security team will be assigned an interrogation from the first wave. In this way, all suspects in the first group will be immediately interrogated upon arrival at the motel instead of first going to a waiting room. In the eight- or ten-member team referred to earlier, primary interrogation assignments would be given to the three strongest interrogators, who are able to work very rapidly. These men would be concerned with what is referred to as a "quick breakthrough"—resulting in a preliminary hand-written statement. Several other interrogators might have the primary assignment as an expansion interrogator—that is, someone who can take over the interrogation after the breakthrough and expand the case more completely in keeping with facts already developed. The polygraph examiner's primary duty, of course, will ultimately be to administer all polygraph examinations. In the meantime, he might very well serve as a breakthrough interrogator or expansion interrogator. Another member of the team would be assigned the responsibility of assisting the secretary in providing food at mealtimes, making searches, recovering evidence, and processing and storing contraband.

Objective. During the briefing the security chief should be sure to stress the primary objective of the closing, which will normally be to develop sufficient evidence to justify a discharge. In certain cases, it may also be possible that the primary objective could be prosecution for one reason or another, and this of course should be stressed to the team so that the case is built accordingly.

Special Instructions. Specific instructions should be given as to policy regarding phone calls and food or beverage for the suspects. Each member of the security team should be told to stress the fact that the suspects are at the motel on company time, under the same conditions as if they were on the job. They should also stress the fact that if it becomes necessary to go beyond the normal quitting time, this will be on a voluntary basis and the suspects will be paid at the rate of time-and-a-half. In this regard, it is imperative that the secretary keep a detailed log as to the arrival and departure times of all suspects so that exact pay times can be computed later. The suspects must also be told that they are not there under arrest. All security personnel should readily identify themselves as security representatives for the company. It is extremely important that it be made very clear to all suspects that there are no policemen present.

Monitoring the Waiting Room. The team should also be made aware of the conditions which will prevail in the waiting room as the day wears on. The waiting room should be monitored constantly by some member of the security staff who is not otherwise occupied at the time. The reason for the monitoring is not so much to act as a guard but simply to psychologically deter the suspects from discussing the case.

Protecting Investigative Data. Admonitions should also be given to the interrogators on the security of their investigative summaries and clipboards which contain the individual statements and related papers for each suspect. They should be told that under no circumstances are they ever to leave an interrogation room containing a suspect and not carry along the aforementioned articles.

Firearms. If certain members of the security team are to be armed, then the proper admonition must be given concerning the display of any firearms. Under no circumstances should a firearm worn by a security person be visible to an employee suspect. In other words, if firearms are necessary for the recovery of merchandise, then jackets are to be worn and kept buttoned at all times. Of course, firearms should never be carried in violation of any state or local laws.

Director's Authority. During the briefings, the investigation director must make it clear that he and he alone will be the boss of the operation. Accordingly, all press and police relations must be handled by the director and no one else.

"Pep Talk." The briefing can usually end with a psychological preparation of the investigation team: a "pep talk" pointing to the anticipated success of the operation and also emphasizing that it will be a team effort and no "grandstanding" by individual interrogators will be appreciated.

The Control Sheet

The interrogators and other security team members should also be instructed on how to report their progress through the day and how to receive new assignments as the day progresses. This is normally done through the use of a control sheet which is maintained by the director of the operation. Experience has shown that in processing large numbers of suspects over a period of days, it is easy to lose track of particular suspects and their various stages of processing at any given time. A control sheet is necessary to maintain proper records. The control sheet contains various columns headed as follows:

NAME
AGE
JOB
YEARS SERVICE
STATEMENTS:
  PRELIMINARY
  EXPANSION
  FORMAL
SELLING
POLYGRAPH
RECOVERIES:
  MERCHANDISE
  CASH
COMPANY ACTION:
  PRELIMINARY
  FINAL
POLICE ACTION
REMARKS

In addition to maintaining control of the investigation and the orderly processing of suspects, the security director ultimately will have to be able to provide top management with detailed information as to the degree of involvement of each particular suspect. Furthermore, in legal action which may follow, such as an arbitration, it may be necessary to reveal to a labor arbitrator the complete scope of the investigation and the degree of culpability from one suspect to another. This is especially true if management ultimately decides to draw a line in an attempt to rehabilitate minor offenders as opposed to major offenders.

Stake-outs

In the advance planning for a closing operation, the security director may wish to provide the capability for stake-outs of certain key suspects' homes during the first evening of the closing. This would hold true especially in cases involving large numbers of suspects where it was not possible to initially process each of the prime suspects on the first day. It must be presumed that many of the employees who have not yet been interrogated will learn of the investigation during the evening of the first day. As a result, experience has shown that where large quantities of stolen merchandise are involved, the contraband may very well be moved from one location to another. By use of a stake-out on key suspects' homes, this diversionary tactic can often be surveilled if not, in fact, intercepted.

Police Assistance

Prior to the advent of the *Miranda* decision (see Chapter 9), police assistance and participation were not only helpful but in some cases highly desirable. Since the Miranda decision by the United States Supreme Court, private security persons are required to operate under the same ground rules as the police department if the police are brought into the case and actively participate with the company security personnel. For this reason, the security director should be very hesitant about making any type of a formal crime report. Under most circumstances, the police should not be involved in the case, as to do so would only hamper the efforts of the company's security personnel and greatly complicate the whole procedure. If any police assistance is needed, it should be unofficial in nature and only advisory in form. Active participation by the police, if any, should come at the conclusion of the case and then only in those cases where there is a clear-cut desire to initiate prosecution.

Bringing in the First Group of Suspects

Because the success of the closing operation depends upon maintaining secrecy throughout the first day, the method of getting the suspects to the motel is critical. The most desirable way is for a trusted member of first line supervision to round up the first four or five prime suspects and simply ask them to accompany him in his automobile to a meeting with a higher-ranking executive. The supervisor should not be aware himself that an investigation is about to commence; he should simply be given instructions over the telephone by the higher-ranking executive and told whom he should bring to the meeting. The supervisor can then be briefed at the motel by the security director as to the role he should play with succeeding waves of suspects. Another pretext that has been used successfully is a request for the suspects to accompany the supervisor to a particular motel to aid in setting up merchandising displays.

In some rare cases, complicated by militant union members, it may not be possible for a supervisor to attempt to bring out the suspects himself. This would be especially true where there was reason to believe that one or more of the suspects might be armed, might be in possession of drugs, or would otherwise be so "street-wise" as to see through the pretext. In this case, it would be much simpler to have the supervisor or personnel department summon a particular suspect (only one at a time) to the front of the plant where several security representatives can identify themselves and request the suspect to accompany them. Experience has shown that where this is done properly and with great skill, refusal on the part of the suspect is unlikely. In the rare event of a refusal, however, the security personnel should be instructed to immediately take the suspect into a private, secluded office and start the interrogation at the plant. This procedure would not be desirable but would have to become a final fall-back position in the unlikely event of a refusal.

If the first wave of suspects are brought out by a supervisor, the same precision planning and speed must be employed by security personnel at the motel as would be required in a pick-up at the plant door. Normally, each of the security personnel will be assigned an individual suspect by name. As the car approaches the predesignated spot at one of the motel rooms, each member of the security team calls out the name of his suspect as the group alights from the vehicle. With each person's name being so enunciated, there is no time for discussion between the suspects. In a matter of ten seconds or less, if the operation is handled properly, each suspect should be ensconced in one of the interrogation rooms with his assigned interrogator.

Succeeding Groups of Suspects

Succeeding waves will probably not be as large in number, since there will not be sufficient security personnel on hand to start an immediate interrogation. However, as the day goes on and the situation becomes less critical, the third and fourth waves can often be simply deposited in the waiting room upon their arrival at the motel. By this time, certain suspects from the first and second groups who have already confessed will be in the waiting room. Recovered stolen merchandise may also be present and in the process of being inventoried. This whole atmosphere then works psychologically in favor of the security personnel, and a group psychology seems to take hold. Subsequent confessions are much easier to obtain than those earlier in the day. To confess at this point seems the natural thing to do, as everybody else has been doing the same thing, and a spirit of "cooperation" seems to prevail among those in the waiting room. To the new arrivals, the sight of stolen merchandise being processed is especially devastating and proves to be a powerful mental conditioner for what follows in the interrogation room.

Chapter 9

Interrogations, Confessions, and Evidence Recoveries

Protecting the Undercover Agent

If there is even the remotest possibility of a labor arbitration or a criminal prosecution arising from an internal theft investigation, the security chief should build his case with the view of ultimately winning in that type of legal setting. One of the most effective tools for the company in a labor arbitration, or even a prosecution, is the testimony of an undercover agent. All of the legal arguments regarding interrogation atmosphere, voluntariness of confession, consent to search, etc., pale alongside the testimony of any undercover agent who can state, with eyewitness certainty, that a particular defendant did steal company property.

If the agent can be utilized as a surprise witness, or as a rebuttal witness in a criminal prosecution, the effect of his testimony is overpowering and almost insures a guilty finding. On the other hand, if union lawyers or defense attorneys learn of the existence of an undercover agent, their whole strategy will be based on utilizing the concept of entrapment as a defense. The successful use of entrapment as a defense requires some preparation on the part of the defense attorney, in consultation with the defendant. If the undercover operative is a surprise witness, this necessary preparation time is usually not available, and the defense of entrapment is only a makeshift one at best.

Keeping all of this in mind, the security director must make every effort to impress upon his investigation crew that the existence of an undercover agent is to be concealed at all costs. In writing the investigation summary, any reference to the undercover agent should be made as "another employee." Under no circumstances should interrogators ever use the work of the undercover agent to facilitate the obtaining of a confession. It would be much better to spend a few extra hours' effort in interrogation, or wait for implications from fellow thieves, than to expose the undercover operative.

On rare occasions, it may be necessary to partially expose the undercover agent in an effort to break down the resistance of an offender and obtain a confession. If this does become absolutely necessary and is used only as a last resort, then the agent should be placed in the role of another offending employee who, under great interrogation pressure, has implicated himself and one or more of his fellow workers. In such a case a confrontation would be arranged between the undercover agent and a non-cooperating offender. Otherwise, the undercover agent should be processed routinely along with the other thieves, so that they all gain the impression that he, too, has been caught and will ultimately "lose" his job.

In the trauma of a "bust," the fact that one employee implicates another, even though it is used during a confrontation, is finally regarded as "just one of those things." The confronting employee may ultimately be considered weak of character, and a particular thief may continue to hold a grudge, but at least criminal defense strategy or union legal tactics will not be altered by one employee confronting another and thus obtaining a confession. In contrast, if an offender is confronted by an undercover agent, the confession may be easily obtained, but at that point the groundwork for the defense of entrapment has been laid.

Discretion in Processing Female Suspects

The processing of female suspects in a motel setting during a "bust" dictates prudence on the part of the director. In the author's twenty-three years of experience in large-scale internal theft cases, most of them including women suspects, never has a female made a charge of improper conduct towards any member of the interrogation team. Even though the interrogation is a one-on-one situation conducted by a male interrogator with a female suspect in a motel room, the following precautions will usually suffice to discourage frivolous charges of improper conduct.

Interrogators should be instructed that when they take a female suspect into the interrogation room, the door should be left open three to six inches at all times. Within several minutes of the commencement of the interrogation, the secretary should enter the room and ask the female suspect if she would care for coffee or a soft drink. This action immediately makes clear to the female suspect that there is another female in attendance in the general area. If the suspect affirms her desire for a beverage, the secretary will enter the interrogation room again within the next five or ten minutes with the desired refreshment. On the other hand, if the suspect declines the offer, then the secretary can re-enter the room in about fifteen or twenty minutes and renew the offer of refreshment. Thus, the secretary has easily entered the room on at least two occasions, and it becomes fixed in the suspect's mind that persons may enter or leave the room at will, and that any loud shouting or screaming would be immediately overheard in the outside corridor.

Female suspects in the waiting room present no problem whatsoever, as there are usually other suspects present and the secretary is in attendance.

During the administration of a polygraph test, it is not possible for the motel door to be left ajar, or for the secretary to enter and leave at will. There are other factors, however, that tend to mitigate this potential problem. The female suspect at this point has spent several periods of time with different interrogators in the various rooms in a one-on-one situation, and generally realizes there will be no opportunity for a charge of misconduct. She is likely to expect the same situation in the polygraph examination room as in the interrogation rooms.

The polygraph examiner himself is used to handling female examinees in the course of his regular business. Mentally, he is quite used to being alone with females in confined quarters. Through his training and years of experience on the job, he has learned to conduct himself with such a professional air that the female immediately recognizes his demeanor and attitude as being similar to what she might expect during an examination by her family physician. The fact that a polygraph examiner must come into physical contact with the female in the attachment of the instrument to her body is largely neutralized by the polygraphist's approach towards the subject. This is an approach which must be entirely different from that used by the initial interrogators who were attempting to obtain a confession. Here, the polygraph examiner is working with the female suspect in an effort to verify the truthfulness of her statements.

The Miranda Decision

In *Miranda v Arizona,* a landmark decision in 1966, the United States Supreme Court ruled that when a suspect was taken into police custody, or from the moment that an interviewee became a police suspect, that person must be advised of his constitutional rights. In essence, these rights consist of the advice that any statement made by the suspect can be used against him in criminal court; that he is not required to make any type of statement; that the interrogation must be halted if the suspect so requests; that the suspect may request the presence of an attorney during the proceedings; and that if he is unable to afford his own legal counsel, one will be provided.

The result of the *Miranda* decision was to revolutionize the criminal justice system and, in effect, to reduce the amount of police interrogations and subsequent confessions in the processing of criminal suspects. The decision was meant to govern the activities of sworn police officers and did not include private or corporate security personnel. In other words, *Miranda* is inapplicable to non-custodial questioning by private citizens.[11] This exclusion has been in effect since the original *Miranda* decision, and at least up to the date of publication of this text. Some authorities speculate, however, that eventually the rule may be extended to trained security personnel even though they are not sworn police officers.

An exception to the above guidelines is the case where a security agent might also be a sworn police officer. A number of security agents have been given auxiliary or special police commissions in order to facilitate their work. The existence of such a commission would mean that the security agent would be obliged to give the Miranda warning prior to any interrogation effort. For this reason, many corporate security departments abolished the concept of their agents being special or auxiliary police officers after the *Miranda* decision.

The only other exception to the exclusion would be in the event that the private security agents were to seek and obtain the active assistance of local police officials in the handling of the case. If the local police participate in the case, then their ground rules must be observed by private security personnel. This would include observance of the Miranda warning and any other rules of criminal procedure which police follow. This concept is a powerful argument against inviting the police to participate in a private company's investigative efforts. In effect, it points up the superior freedom enjoyed by security personnel in handling their own cases, compared to the police in handling criminal suspects.

---

[11]Yates v United States. 384 F. 2d 586 (5th Cir. 1967).

Union Representation

In a 6-3 decision, the U.S. Supreme Court has ruled that employees who belong to unions are entitled to union representation during questioning by an employer or the employer's agent.

In two separate but similar cases, *N.L.R.B. v J. Weingarten, Inc.*[12] and *I.L.G.W.U. v Quality Manufacturing Company et al.*[13], the high court ruled that such interrogation, which could result in disciplinary action, entitles the employee to representation if he requests same. Failure to honor such a request "interferes with, restrains and coerces the individual right of the employee to engage in concerted activities for mutual aid or protection."

Without question, the decision is bound to complicate the work of many security officials whose companies are organized by a union. The court did *not* state that representation must be offered in all cases—only if it is desired. Furthermore, the court *did* state that the employer or his agent may elect to terminate the interview if such a request for union representation is made. This option can be pointed out to the employee, as well as the fact that the employee would not gain any of the benefits which might result from such an interview. In other words, the employer may proceed on a course of action with whatever facts are already at hand, and the employee would take his chances with the consequences.

If the interview or questioning proceeds with a shop steward or business agent present, the court affirms the union's role as follows:

> The representative is present to assist the employee, and may attempt to clarify the facts or suggest other employees who may have knowledge of them. The em-

---

[12]N.L.R.B. v J. Weingarten, Inc. 485 F. 2d 1135 (1973).

[13]I.L.G.W.U. v Quality Manufacturing Co. 95 S. Ct. 972 (1975).

ployer, however, is free to insist that he is only interested, at that time, in hearing the employee's own account of the matter under investigation.[14]

It is well established in labor law that an employee has a duty and obligation to cooperate with his employer during an investigation. By the same token, many lawyers feel that a deliberate attempt by a union representative to block or interfere with an investigation would constitute an unfair labor practice on the part of the union.

In a footnote to its decision, the court also makes reference to polygraph tests and the need for union representation. The court is not specific on this issue, but some legal authorities feel that the question of union representation could be satisfied by allowing the representative to be present during a general discussion of the polygraph test and a review of the relevant questions. He should not be allowed to sit in on the examination itself.

Unquestionably, there will be more rulings and decisions growing out of all of this. Also, many unions will have their own interpretation of this decision which will differ from the opinion expressed above. The final decision, of course, will have to be made once again by the courts.

The court's decision on union representation has changed procedures very little in the several intervening years. Generally speaking, most employees are not aware of the court's decision and therefore do not ask for union representation during a major "bust," especially where the elements of surprise and adequate manpower are viable factors. Once the word spreads among employee groups, however, the result of the court's decision will virtually dictate that security directors must develop sufficient evidence beforehand that would enable the case to stand on its own merits in the absence of a confession. Moreover, if such strong evidence is developed beforehand, the interrogator's chances of dissuading the

---

[14] *Ibid.*

employee from his initial request of demanding union representation will be greater. If an employee feels that the company has a solid case against him, and that if he persists in his request for union representation the interview will be discontinued by the employer, resulting in the employee's subsequent discharge, he will be more inclined to wish to discuss the matter privately. He may even hold out some hope in his mind that the employer may allow him a chance for rehabilitation and probation.

An informal survey taken of some key security directors of major corporations indicates that most plan to develop strong investigations and will not attempt any interviews in the presence of the shop steward or business agent of the local union.

Voluntariness and Other Considerations

Without having to contend with the *Miranda* decision, the main issue for the security director to be concerned with on written statements is their voluntariness. In the legal proceedings following the "bust," some defendants will attempt to repudiate their confession with arguments such as: "I was scared and I didn't realize what I was signing"; "I was told if I signed the paper I would not lose my job"; "I was told that the statements would only be used for the insurance company and that nothing would happen to me"; "I was told I would be turned over to the police if I did not cooperate"; "I was kept in a motel room and was not allowed to leave."

The above are just a few of the many and varied stratagems employed by defendants in labor arbitrations or criminal prosecutions in an effort to refute their previously made statements. Knowing all of this, the security director must incorporate into all of the statements certain features that will tend to counter many of the above arguments and make it more difficult for a defendant or his attorney to successfully attack the validity of the confessions. The atmosphere throughout each day of the investigation should be one of informality, friendliness, and one that would tend to uphold human dignity.

For many years the techniques which have been outlined in Chapters Seven and Eight of this text have met the test in lower courts and labor arbitrations successfully. Finally, in *U.S. v Maddox* (1974),[15] the United States Court of Appeals for the Fifth Circuit upheld many of these techniques. The United States Supreme Court declined to accept a further review of this case on a writ of *certiorari*, thus establishing that legal history had been made concerning many of the techniques advocated by this text. In illustration, we quote from the court's own language of this case:

> The voluntariness of K's confession is challenged on appeal. The private investigators arranged for employee suspects to be questioned at a local motel, concerning the missing merchandise. The employees, still on the company payroll and not under arrest, were maintained at company expense in a casual atmosphere. A TV room, sandwiches, and snacks were provided. The employees received their normal pay, including overtime during the questioning. Their freedom to leave was equivalent to that while "on the job." It was in this context that K confessed his activities.[16]

Outside Suspects

In many cases persons who are not directly employed by the company may be involved in internal theft rings, and it becomes necessary to consider a method of prosecuting such outsiders. Realizing that he does not have the same degree of psychological control over an outsider as over an employee, the security director knows that an outsider's presence at the motel for processing must be completely voluntary. Experience has shown that many outside suspects will cooperate if approached during the first day of the "bust" by several agents

---

[15]VOL. 492 F 2d. 73-2611 (1974).

[16]*Ibid.*

who identify themselves as security representatives of the company. The majority of these people usually are willing to accompany the agents back to the motel for questioning. In trying to analyze their motives for compliance, one can only assume that, knowing they are deeply involved in the theft, they reach the conclusion that they would much rather take their chances with company personnel than with the police. Of course, security agents must be careful not to use the threat of police involvement as a means of gaining their cooperation.

Once at the motel, the outside suspect must again be made aware of the fact that he is not under arrest and is free to leave if he so chooses. Other than that, he should be processed like any other company employee, but the processing should be expedited because of the fact that he is there on his own time. Following completion of his processing, the outside suspect should be returned to his home or place of business. If he is to be prosecuted, his individual case should be included with all of the others that ultimately are turned over to the police or district attorney following completion of the case.

Occasionally, it will be determined that an outsider is the employee of another company. In that case, it may even be possible to gain the cooperation of the other company's security department in providing for interrogation facilities on the other company's premises.

The Preliminary Statement

As discussed in Chapter Eight, the initial interrogation of the suspect is designed to obtain a quick "breakthrough." This means that the primary objective is to obtain a written statement of any significant theft which is initially admitted by the suspect. Experience has shown that if a case is properly prepared beforehand, and sufficient information has been developed, as contained in the investigation summary, a preliminary statement can be obtained from most suspects within the first fifteen to thirty minutes of interrogation time. The main objective here should be a statement that will sustain a discharge, if challenged in a labor arbitration. If at all possible, the statement should be in the suspect's own handwriting, and should contain sufficient information to hold up under attack on *voir dire.* An example of such a preliminary statement, which has been upheld throughout the years in various legal proceedings, is shown in Figure 9-1.

The Expansion Statement

Following the preliminary statement and a brief pause for refreshments in the waiting room, a second interrogator will normally take the suspect to another interrogation room in an effort to expand the information already obtained. This will be recorded in a so-called addendum statement. Many times, the approach used by the initial interrogator for a quick breakthrough will almost dictate the necessity for a new interrogation face—one who can utilize an entirely different approach which is necessary to expanding the degree of involvement.

The purpose of the expansion statement is to develop knowledge of coworkers' involvement, collusion, methods of operation (M.O.), location, etc. The expansion interrogator may very well become more aggressive in his interrogation than did the "breakthrough" interrogator. There is less chance of losing the subject during expansion than during the initial interrogation.

In any addendum statement, the time and date should be noted and reference made to the earlier statement made in the initial interrogation. Here again, the statement should be in the subject's own handwriting if at all possible. It is convenient to use an NCR-type triplicate paper for the preliminary and any addendum statements. This type of paper lends itself easily to handwritten statements and produces its own copies, thus eliminating the necessity for carbon paper. All that is required is a firm writing surface and the use of a ball-point pen. An example of an expansion or addendum statement is shown in Figure 9-2.

The Narrative Statement

Although the narrative statement is very popular with insurance investigators and with certain law enforcement agencies,

Baltimore, Md.
Jan. 9, 1978

My name is John Smith. I am 40 years old and live at 1234 Main Street, Baltimore. I am employed as a shipping assistant at Acme Products 567 South Street, and have worked for the company for 1½ years.

During the last 6 months of my employment, I have taken without permission or payment property and merchandise belonging to Acme as follows: 7 full rolls of piece goods, 2 partial rolls of piece goods, one hand truck, and about 6 shirts. I took the shirts out at lunch time by hiding them under my clothes. My friend Bill Jones who is a truck driver helped me get the piece goods out.

This statement is true and voluntary. No promises or threats have been made to me. I knew this was wrong but ask the company to give me another chance to prove myself.

John Smith
1234 Main St.
Baltimore, Md.
10 AM

WITNESSED:
Kirk Barefoot
Rose M. Reilmann

Figure 9-1. Preliminary statement.

Jan 9, 1975
11:45AM

My name is John Smith. I wish to add to the statement which I made earlier today. After discussing the thefts with Mr. Harris, I realize I did not tell everything in the first statement.

Altogether Bill Jones and I have been stealing rolls of piece goods for almost one year. On Wednesdays I am left in charge of the shipping and receiving doors during lunch time. Bill usually gets there with his truck about 12:45 and we usually put something on. Bill gives me half of what he gets for the cloth. I guess I have received about $1,000.– for my share. Bill says he sells the cloth to his cousin who lives in Pennsylvania. We have taken somewhere between 30 and 40 full or partial rolls.

I make this second statement so that the truth will be out. I want to cooperate and I know I will probably loose my job. I am being well treated.

WITNESSED: [illegible]
[illegible]

John Smith

Figure 9-2. Expansion statement.

its use is advocated by the author primarily under two circumstances. In a major investigation, where a preliminary statement has been obtained, and this in turn has been followed by one or more more addendum statements, it may be desirable to finish the processing of an individual by taking the narrative statement and attempting to incorporate into it all of the various points developed, rather than having them appear in different statements. In other words, the narrative statement could be considered a summary of a suspect's involvement in the case. This approach should primarily be used where the director feels that no prosecution, or any other type of legal activity, criminal or civil, is contemplated. Also, the narrative statement is generally not recommended for use in building a bonding claim.

A further use of the narrative statement would be when management must act in a situation where an employee has been apprehended and no professional security personnel are quickly available to take a statement. Here, it is to the advantage of management to get as many of the facts as possible into one statement.

The main objection to a narrative statement is the fact that it is more self-serving than other types of statements. Unlike the preliminary statement, where only the first and last paragraphs are conceded to be dictated, the narrative statement is usually dictated in its entirety, and it is the interrogator's attempt to tell the suspect's story for him. In such an attempt, the language used in the statement is almost never exactly the type that might actually be used by the suspect. The statement itself, of course, tells a neat, concise story, which clearly benefits the investigator rather than the subject. At no point in the narrative statement does the subject clearly have an opportunity to say anything which he may wish to include in such a statement. A sample narrative statement is reproduced in Figure 9-3.

The Formal Statement

Although more time-consuming and involved, the formal statement is by far the strongest type of statement which an in-

SPECIMEN OF
NARRATIVE STATEMENT

August 1, 1974

"I, John Doe, age 23, am employed by ______________ Company, Newark, New Jersey, as a shipping sorter. I have worked for ________________ Company since June, 1962.

Today, at about 1:30 PM, I took an empty box and went to the ________________ section, where I removed the following items from section 8 (list):

____

____

____

____

I fully realized that this was the property of my employer and I realized that I was stealing because I intended not to pay for them but to sell them to my cousin who lives near me. I carried the box down to the shipping room where I hid it until the end of the day. No one in authority gave me permission to bring the box down, as I came down the back stairs when no one was looking.

After quitting at 5:00 PM, I went to the rear of the building and entered the shipping room where I picked up the box. Mr. Williams, the Foreman, caught me just as I jumped down off the rear dock.

I knew I was doing wrong and stealing, and now want to do right by the company by telling the truth. I put my initials on all the merchandise I stole, and it has been sealed up by Mr. Williams, in my presence, for safekeeping. I also wrote about the items on the outside of the bag and signed my name.

I have made this statement of my own free will and voluntarily. No promises or threats were made to me, and I feel that I have been treated fairly. I made this statement so that the truth would be known."

Signed:

John Doe
Newark, New Jersey

Witnessed:

Mr. J. B. Williams, Foreman
Miss Jean Smith, Secretary

Figure 9-3. Sample narrative statement.

vestigator can obtain. It is a statement which, by its very nature, forces the suspect to utilize his own language and his own descriptions of events which have taken place. By virtue of the fact that it is first taken down in shorthand by a stenographer, it insures the presence of another witness throughout the actual taking of the statement. If structured properly, the statement can not only overcome any questions as to voluntariness, but can also give sufficient detail upon which to base specific criminal charges of larceny. This type of statement has withstood the challenge of time in the many court jurisdictions throughout the United States. To put it simply, if the statement is properly taken, there is no way the suspect can repudiate it at a later date.

Figure 9-4 is a sample prototype for a formal statement designed to cover larceny cases. In embezzlement cases, or any other case involving various forms of documentation, the only difference would be that each piece of documentation should be read into the statement. That is, where certain records have been altered or forged, each individual record should be presented to the suspect and identified in the statement, and the suspect should be asked to examine the document and initial it. This is then acknowledged in the statement as one of the documents utilized in the crime.

In the event of a case involving collusion of two or more suspects, it is sometimes desirable to have the second suspect present in the room when the statement is being taken, and to indicate in the heading of the statement that the additional suspect is present. Furthermore, it is often desirable to have the second suspect sign as a witness on the statement of the first suspect, if possible. This whole situation is then reversed when taking the second suspect's formal statement. The fact that various members of a theft ring or collusive theft are present and witnessing each other's statement is a powerful binder on the whole case.

Searches, Seizures and Recoveries

During the taking of a preliminary or addendum statement, a suspect will often indicate that he has placed stolen property

STATEMENT

Statement made by William G. Doe, employee of____________________________ Company, in the office of Richard Long, Plant Manager, at 123 Washington Road, Baltimore, Maryland. This statement was taken at 3.00 PM, January 5, 1974, in the presence of Richard Long, Plant Manager, and Mary White, stenographer. Questions, unless otherwise indicated, are by Thomas Jones, Security Officer for the Company.

Q. What is your name and address?

A.

Q. What is your age, date of birth, place of birth?

A.

Q. Where are you employed, in what capacity, and for how long?

A.

Q. Can you tell us the name of the grammar school you attended?

A.

Q. What was your mother's maiden name?

A.

Q. Before proceeding further, do you realize that you do not have to make this statement - it could be used against you?

A.

Q. Why do you wish to make this statement?

A.

Q. Do you realize that you can refuse to answer any of the questions in this statement?

A.

Q. In connection with your business as a________________________________ for______________________________Company, do you have occasion to handle items of merchandise which are the property of your employer?

A.

Q. In connection with your handling this merchandise, have you ever stolen and removed any of it from the____________________________?

A.

Q. When you took this merchandise, did anyone in authority give you permission to take it?

A.

Q. In not having permission to take the merchandise, nor paying for it, did you realize you were stealing and in effect committing a larceny?

A.

Q. Bill, can you tell us to the best of your memory, when you first stole and removed merchandise from the________________________?

A.

Q. Can you tell us the most recent occasion when you have stolen merchandise from_______________________? (Get details of date, time, method of removal, etc.)

A.

Q. What is the least expensive item you have stolen from____________________?

A.

Q. What is the most expensive item you have stolen from_____________________? (Get as many details as possible).

A.

Q. What is the most merchandise you have ever stolen and removed from the __________________________ at any one time?

A.

MR. LONG NOW QUESTIONING:

Q. Can you remember on what day or in what month this occurred?

A.

Q. Bill, can you tell us to the best of your memory, the total amount of merchandise that you have stolen from__________________________?

A.

Q. On what basis do you arrive at the above figure? (Subject's answer should be at rate of so much per day, week, month or year).

A.

Q. Can you give us examples of merchandise which you have stolen which are included in these figures?

A.

Q. Can you tell us why you stole these items? For what purpose did you intend to use them? (If the items were for resale, the interrogator should get the name and address of the receiver, what items were delivered, where the items were dropped off, how much money was received, dates of deliveries, method of ordering, etc.).

A.

MR. JONES AGAIN QUESTIONING:

Q. Bill, in stealing these items, how did you manage to get them out of the_______________________?

A.

Q. Since you have worked here, have you ever helped or participated in stealing with other employees?

A.

Q. Can you tell us who these employees are and just how you were involved with each one? (Here again, the interrogator should get as many details as possible - that is, who, what, when, where, why, how, etc.).

A.

Q. With regard to all of the merchandise now in your home, car, garage, etc., would you be willing to turn it over to us personally?

A.

Q. Bill, have you made this statement voluntarily and of your own free will?

A.

Q. Have any promises of reward or immunity been made to you?

A.

Q. Has anyone forced or threatened you to make this statement?

A.

Q. Can you tell us if you feel you have been abused or if you have been treated fairly?

A.

Q. Is there anything else you wish to tell us that has to do with this matter? (Interrogator should repeat this question until a "No" answer if forthcoming).

A.

Q. If, after reading this statement, and making any necessary corrections, additions, or deletions, will you be willing to sign it?

A.

"I have read this statement, and understand its content fully. The statement contains________pages, and I have written my initials on each page in addition to signing my name on this page."

Date:______________________________ Signature:__________________________

Witnessed: Address in full:

______________________________ ______________________________

______________________________ ______________________________

____

Figure 9-4. Sample formal statement for larceny cases.

in his home or in some other area. If he has not volunteered this information, then it should be explored by the interrogator and an attempt made to determine the existence of such stolen contraband which could be recovered. At this point, after he has made one or more handwritten statements of guilt, there seems to be little problem in obtaining the subject's consent to a voluntary search of his home or some other area. For this purpose, experience has shown that a Consent to Search form, as shown in Figure 9-5, has proved more than adequate.

In the example shown, it will be noted that there is a portion on the bottom of the form for an additional consent to be given by the spouse. In some cases, the investigator may run into a refusal to cooperate on the part of the spouse. This should not deter the search, as long as one of the spouses has consented to the search.

In most searches of this type, the investigator is almost totally dependent upon the suspect to identify the stolen merchandise for him so that its recovery can be effected. During the trip to the house, and during the search, the suspect should be reminded that the total procedure includes a polygraph test. He should be told several times that one of the questions will cover his cooperation in turning over all stolen property during the search. The thought of the polygraph test acts as a great mental prodder in encouraging the suspect to release all of the stolen goods.

In other words, the search is not a true search in the sense of a police-type operation. Nothing is going to be removed from the premises except that which is identified by the suspect as having been stolen. The only exception to this would be merchandise in an original container which would clearly indicate that it was stolen, and where the container itself would rule out the possibility of the merchandise ever having been obtained legally. In this case, without the suspect's cooperation, the investigator should make a seizure of the item and, of course, give the suspect a receipt.

The investigator should not attempt to search the home of a

CONSENT TO SEARCH

I, ______________________________, age ______, admit that merchandise stolen from ________________________________ is presently stored at __________

____.
(Description and Location)

I hereby consent to take ____
(Names of persons making search)

____

to ____________________________________, and agree to return all of such stolen
(Description and Location)

merchandise to the above named individuals, after identifying the stolen items by placing my initials thereon.

I, certify that the above consent has been freely and voluntarily given by me and that no threats or promises have been made to me.

I further agree that the above named individuals may enter the above described premises to assist me in locating the stolen merchandise, and that I shall make no claim whatsoever against the above named persons, (Company Name), its officers and employees, in connection with the entry and search of the above premises and the return of the stolen merchandise.

Date: ____________________

Witnesses: Signature: ____________________________

____

____

Consent of Spouse, or Co-owner, or Proprietor of Premises:

I, ____, hereby consent to the entry of
(Name of person and relationship to employee or premises)

the above described premises by the above named individuals.

Date: ____________________

Witnesses: Signature: ____________________________

____

____

APA - 1973 SAMPLE

Figure 9-5. Consent to search form.

minor if the parents are present and refuse cooperation. If the parents are not home, then the investigator should confine his search to the minor's room in the home. Even in so doing, there is a certain risk of eventual lawsuit on the part of the parents, but many security directors might well consider this an acceptable risk.

Although many security agencies have a certain expertise in interrogation and the taking of statements, a great number of them hesitate to pursue a case beyond that point, and in reality make little or no effort to effect a recovery of merchandise. Some security people are content simply to allow the suspect to "bring in the merchandise the next day." With this approach, all of the merchandise may not be recovered, but it probably does reduce somewhat the risk of a lawsuit. It would also reduce the risk of bodily injury to the investigator. Experience has shown that in an operation of this type, the most dangerous situation which can present itself occurs during the searching of a suspect's home. Here, the chances of finding hidden weapons or belligerent relatives are all greatly increased. Unless an investigator is armed it is not advisable for only one person to conduct house searches.

The secretary should also play a key role in any subsequent search of a female suspect's home. Under no circumstances should a male investigator take a female suspect to her home alone, even though she has signed a Consent to Search form. She should be accompanied throughout the entire procedure by the secretary, whose presence, once again, greatly deters the possibility of a later charge of impropriety against the investigator.

Once the stolen merchandise is recovered, it should be returned to the motel and processed in the waiting room. This can easily be done in front of the other suspects, and serves to act as a powerful pre-conditioner to their own interrogation or house searches. In processing the merchandise, the investigator can enlist the suspect's assistance in marking each item for identification. The entire inventory and marking of the items should be recorded on a Property Recovery sheet (Figure 9-6), which ultimately forms a permanent record as to

PROPERTY RECOVERY SHEET Page________ of ________

The following is a list of property stolen by ______________________________
(subject's name)

from ______________________________ located at ______________________________
(company name) (company address)

and surrendered to/seized by ______________________________, ____________________
(circle one, cross out other) (name) (title)

at ____
(describe fully where recovery occurred)

at ____________________ on ______________________________.
(time) (date)

SUBJECT: ______________________________ POSITION: ______________________________

ADDRESS: ______________________________ CITY: ____________________ STATE: ____________________

WITNESSES TO SEARCH OR RECOVERY: (give names and addresses)

____

____

RECOVERY MADE BY:

- ☐ Search Warrant
- ☐ Consent to Search
- ☐ Voluntary Surrender
- ☐ ____________________ (other)

IDENTIFICATION MADE BY:

- ☐ Subject ____________________ (name)
- ☐ Sec. Off. ____________________
- ☐ Management ____________________
- ☐ ________ (other) ____________________

QUAN.	SIZE	AREA OF RECOVERY	DESCRIPTION OF ITEM/MERCHANDISE	UNIT PRICE	VALUE

"I have personally identified the above items/merchandise as being the property of ____________________
(company name)

and certify that this is an accurate record of this recovery. I have completed this form voluntarily and no promises or threats have been made to me." **WITNESS TO SIGNATURE:**

____________________ ____________________
(signature)

____________________ ____________________
(date and time)

The above items/merchandise have been marked in the following manner:

____
(describe fully how items/merchandise were marked)

____
(signature of security officer)

Figure 9-6. Property recovery sheet.

how and under what circumstances the merchandise was recovered. Months or even years later it may become extremely important to know all of the details surrounding the recovery of the merchandise, and especially any witnesses present. By proper use of the Property Recovery sheet, a permanent record is obtained which can prove invaluable in any criminal proceedings which may later develop.

Chapter 10

Role of the Polygraph And Disposition of Offenders

THE POLYGRAPH AND INVESTIGATION

Supportive Role

The ability to use polygraph examinations in conjunction with the overall efforts of the "bust" is highly desirable. It makes for a much more complete investigation. It is through the use of the polygraph that the true perspective of the case can be fully developed and management can be made aware of the full extent of their losses. The inability to utilize polygraph tests in this type of setting is not fatal to the company's investigative objectives, however. The polygraph's role is an ancillary one rather than a technique utilized in a first-line offensive.

The following case history illustrates why the polygraph should have a supportive role rather than carrying the brunt of the investigation.

Madison, Wisconsin . . . One of the top polygraphists in the country was retained by a client to investigate inventory shortages in its branch at Madison, Wisconsin. The principal location of the client company was non-union, and over a period of years the client had enjoyed the ability to polygraph employees at will. In the Madison branch, however, the operation was organized by a Teamsters local.

Upon arrival at the Madison facility, the polygraph examiner, working alone and without any definite plan of action in mind, proceeded to set up his instrument in the general offices. After a brief discussion with local officials, the first suspect was summoned to the office and introduced to the polygraphist. Without too much difficulty, the employee voluntarily submitted to an examination. Several minor admissions of theft were obtained, along with the names of several other employees who were implicated in additional thefts.

After testing and releasing the first employee, the polygraphist summoned one of the others implicated. About ten minutes into the pre-test interview, there came a loud knocking at the door of the examination room. There stood the union shop steward demanding to know what was going on and insisting on sitting in on the proceedings.

Within an hour, the Teamsters business agent had arrived and held a meeting with the remaining employees of the facility. At this point, the polygraphist realized that he would not be able to interview any of the employees without the shop steward being present. Furthermore, none of the employees would agree to a polygraph examination, nor was there any hope of obtaining any written admissions of guilt. Faced with this kind of opposition, the polygraphist had no alternative but to pack up his instrument and leave for home.

As a result of the company's failure in its attempt to investigate the internal thefts, the thieves not only went undiscovered, but the inventory variances continued.

This case demonstrates what usually happens when a polygraph examiner accepts such an assignment from one of his clients. Similar situations have been experienced over the country in recent years by countless polygraphists who have attempted to take on the investigation of major theft cases in union facilities, using their polygraph as the number one investigative tool.

Even in a non-union facility, over-reliance on the polygraph in investigation can bring undesirable results. It becomes readily apparent to most of the employees that the company does not have any facts at hand, or only a few at most, and that the sole reason for investigation by polygraph is that the company is seeking answers it does not already have. Such an investigation becomes somewhat of a witch-hunt in the sense that people who are not involved in the thefts at all are subjected to the polygraph examinations along with the dishonest employees. This often has a detrimental effect on morale, tending to demoralize the honest, loyal employees in the work force. Furthermore, the dishonest ones many times are able to gain moral support from the honest segment, and in a number of cases have been instrumental in causing a particular non-union company to become organized by a labor union. Frequently, dishonest employees feel that the introduction of a labor union into the industrial setting will serve as a shield against management's efforts to discover the true identity of thieving workers.

It should be pointed out that the regular and routine administration of periodic polygraph examinations is a different matter. In such cases, the employee is usually made aware at the time of employment that he or she will be given an initial polygraph test, followed by periodic examinations which could range in frequency from every six months to every one or two years. In this type of situation, the employee is mentally conditioned to the fact that polygraph examinations are a part of his work routine, and he tends to accept such a program, albeit not enthusiastically. Surveys have found that the vast majority of working people have absolutely no objection to pre-employment polygraph examinations and generally will accept periodic examinations if they are made a pre-condition of the job. The real resentment comes about when the workers are summarily given polygraph examinations in order to clear up real or imagined shortages due to theft.

For all of these reasons, the polygraph's proper role in a large "bust" situation is that of a supportive investigative tool,

VOLUNTARY CONSENT TO POLYGRAPH EXAMINATION

I______________________________, age________, of my own free will and without duress, agree, in connection with my employment by ______________________________ to submit to polygraph tests. I understand that the term polygraph test is more commonly known as a "lie detection test". I further authorize the attachment of the polygraph recording instruments to my person. I agree that the results of such tests be made known to______________________________, and that said results become part of my personnel files.

I hereby waive and release any and all claims and causes of action of every kind whatsoever against______________________________, its officers and employees, and any person, firm or corporation engaged by ______________________________, in connection with the taking of such polygraph tests or in conducting any investigation concerning my background, which I may now or in the future have, arising out of or in connection with the aforesaid polygraph tests, or investigative procedures.

I have carefully read all of the foregoing and fully understand its contents.

______________________	______________________________
(Date signed)	(Employee's signature - Use ink)

Signed in the presence of______________________________

Figure 10-1. Voluntary consent form for polygraph examination.

one that must be integrated with the other investigative techniques utilized by the security director.

Proper Polygraph Applications

When utilized properly as a support tool in the "bust" situation, the polygraph technique has the principal role of verifying admissions, true total amounts of goods stolen, the identification of co-offenders, and the location of stolen goods. In the verification of admissions and true total amounts, the best time for a polygraph examination would be following a house search but preceding a formal statement. In other words, no formal statement should be attempted unless the security director feels certain that all of the information that can be developed on a particular suspect has been developed. There is one and only one way that this can be done, and that is through polygraph technique.

As a part of the general processing of an offender in a large group, it is customary that he be asked to sign a polygraph consent form some time after making an initial preliminary statement and prior to the house search. A sample of such a consent or waiver form is presented in Figure 10-1. On rare occasions, an offender may refuse to sign a polygraph consent form but will make a verbal statement that he is agreeable to taking the test. If this situation should be encountered, the security director can easily call the suspect's bluff by having him indicate in front of at least two witnesses his willingness to take the test. Many times, such an offender is merely playing a game and will ultimately refuse to take the test, but in isolated cases he will go forward. If this be the case, a memorandum should be typed up for the signatures of the two witnesses indicating the time, date, and place where they overheard the verbal consent and the circumstances under which it was given.

The Verification Test

In any large "bust" situation, the following relevant questions should be included in the polygraph examination:

1) Do you know anyone stealing from ______ company? (optional)
2) Have you ever worked in collusion with anyone to steal merchandise from ______ company?
3) Have you stolen (any) more than ____ dollars in property from ______ company?
4) Have you ever sold any stolen ______ company property?
5) Do you know where any stolen ______ company property is located?

The above test questions are those most commonly utilized in a large-scale industrial theft investigation. Occasionally, the security director and his polygraph examiner may wish to resort to other more specialized tests.

Peak of Tension (P.O.T.) Tests

On occasion, an employee suspect may be encountered who will make some minor admissions of theft, but, for one reason or another, refuses to expand his admissions and give a true amount. Where the security director has reason to feel that this is the case, he may request his polygraphist to utilize a peak of tension test designed to pinpoint the total amounts of theft.

A basic example might be where a suspect has made an admission of approximately ten dollars; the question range would start with one dollar, and then five dollars, and continue on up into the hundreds or even thousands, as the case may be. Most polygraphists then prefer to come back down the scale in reverse order. If the examinee is properly prepared psychologically, the polygraphist will usually be able to pinpoint the general area of the true amount of his thefts. Although this certainly is not conclusive from the standpoint of winning an arbitration or a prosecution, it has a very beneficial effect on the suspect and is a great aid to an interrogator who may wish to attempt further questioning.

Basically the same type of test, although of a more probing

nature, can be used to identify co-offenders, and also the location of hidden contraband.

Properly used, then, the polygraph can be of great assistance in the overall investigation. It also plays an important role in the disposition of offenders and rehabilitation of employees, if this is to become a factor.

DISPOSITION OF OFFENDERS

The Decision to Retain or Discharge

In any large theft investigation, management is often faced with a final decision as to who should be separated from the payroll and who should be retained. If the investigation was widespread, identifying many of the employees as engaging in dishonest acts, management may be forced to make a business decision based on a desire to keep the facility in operation. In many cases, it would be foolhardy to discharge each and every admitted thief, as there would not be sufficient trained personnel remaining to keep the company's business going. Therefore, the decision comes down to where to draw the line.

If there is a union in the picture, management must be circumspect in drawing such a line so that they do not deviate from their own standards and are consistent in applying them uniformly. If the facility is not unionized, management has the prerogative to discharge whomever it wishes and, conversely, to retain any persons deserving of such retention.

By studying the security director's control sheet, management should be able to determine at what point a reasonable cut-off line can be drawn. It is not practicable to attempt to suggest any definite dollar amount of theft that should be grounds for discharge, as all amounts must be relative from one case to another. That is, in some cases amounts of admitted theft in the area of two to four hundred dollars might very well be considered minor, whereas in other investigations, minor amounts might very well be considered those less than fifty dollars.

The security director, however, should attempt to dissuade management from retaining any persons who have been unable to verify their own admissions through polygraph; that is, those whose confessions could not be verified as being complete. Also to be excluded from any consideration of rehabilitation would be those persons admitting the sale of company property, regardless of the amount involved. To put it another way, only those offending employees who have admitted minor amounts of pilferage for their own personal or their family's use, who have not engaged in selling stolen merchandise, and who have been able to verify that their admission is complete through polygraph examination, should be considered worthy of retention. The company must be prepared, however, to clearly state what those conditions are that influence the discharge of the other offenders. If they are consistent in adhering to the standards that they themselves set up, the company should be successful in gaining the support of any labor arbitrator.

In regard to those who are retained, it is beneficial to be able to show an arbitrator that such retention has its own conditions—namely restitution by payroll deduction and an agreement for re-polygraphing at some future time. A suggested form to be used in re-polygraphing is shown in Figure 10-2.

Larceny Charges

Regardless of any desire on the part of management to prosecute, many larceny cases never find their way into the criminal justice system. When they do, the most common charge is larceny, but that very charge carries with it certain complications which reduce, overall, the number of thefts referred to the police or district attorney. The main complication arising in a larceny case is that of the proof of the *corpus delecti*. This literally means the body of the crime. In any criminal proceedings, it is absolutely essential that the prosecution first prove that a crime was, in fact, committed. The next step is to prove who did it. For this reason, confessions, in and of themselves, are insufficient to win criminal prosecution.

Consent to Re-Polygraph

As a result of the recent security investigation, I confessed to taking or removing certain items of merchandise which were the property of   However, has offered to continue my employment on a probationary basis with the express understanding that any recurrence of stealing on my part will result in immediate discharge. I further agree to take polygraph tests in the future at the request of the company.

I thoroughly understand the conditions set forth by the company as outlined above, and I fully consent to these conditions without any qualifications whatsoever.

DATE: ____________________   SIGNED: ______________________________

Figure 10-2. Suggested consent form for re-polygraphing.

The burden of proof of the *corpus delecti* rests upon the prosecution and not upon the defense. An inventory shortage is insufficient to establish a *corpus delecti,* although it can be a part of the ingredients of the *corpus delecti,* if other facts are there to support it. The *corpus delecti* can often be shown in a number of ways, such as breaking and entering, eyewitness testimony, or even the recovery of stolen company property.

Identifying Stolen Property

If recovered company property is to be used in establishing the *corpus delecti,* then the prosecution has the burden of showing that the property in question can be positively identified as stolen property, and that the defendant could not have come into possession of such property in any legal manner. Once again, the burden of proving this rests upon the the prosecution and not the defense. In other words, the defendant does not have to prove that he came into possession of the property legally, but rather the prosecution has to show that he could not possibly have come into possession of the property legally.

The following case history illustrates the painstaking efforts that can go into proving the *corpus delecti*.

Long Island City, New York . . . . After many months of surveillance of a liquor warehouse, the investigation team concluded that a shop steward on the night shift was utilizing his personal automobile to remove cases of Scotch whisky from the warehouse. Surveillance had established a regular pattern of deliveries to various customers in the borough of Brooklyn. In order to be able to prove that there was no chance that the subject could have purchased the cases of whisky, the investigators were forced to go to greath lengths to make the problem of the *corpus delecti* airtight.

Inasmuch as it had been planned to intercept the shop steward on a Monday morning while he was making his deliveries of whisky, the investigators spent all of Sunday inside the warehouse, marking certain brands of Scotch whisky so that positive identification could be made the following morning if the apprehension were successful. This task was enormous, as there were hundreds and hundreds of cases to be marked in such as way as not to alert the shop steward.

As events later proved, when the steward and the night crew reported at midnight on Sunday night for work, the shop steward loaded six of the marked cases into the trunk of his automobile which he had parked inside the building. When the apprehension was made, there was no problem in identifying the six cases and eliminating even the remote possibility that he could have purchased the six cases on his way to work that evening.

Other Charges

Occasionally, when the security manager finds that proving the *corpus delecti* is an insurmountable task, he may find that he is able to prosecute on the basis of conspiracy. In a conspiracy, it is not necessary to prove that a crime was committed. It is only necessary to prove that two or more persons conspired to commit the crime and that at least one overt act was performed towards that commission. In a theft ring situation, this often can be proved through the testimony of one of the conspirators, or possibly even an undercover agent.

In some states, the charge of embezzlement is applicable even though it may not involve a company accountant or the actual financial records of the firm. Basically, embezzlement is a charge that applies when someone who is entrusted with the custody of the goods absconds with same, as opposed to a larcenous person who may merely have access to the stolen goods. In a true embezzlement, the proof of the *corpus delecti* is usually much easier than in a larceny case, as the altered financial records of the company become, in fact, the body of the crime.

Other charges that may be available to the security director in order to make his case are drug violations and, of course, gambling charges. Here, the recovery of illicit drugs or gambling paraphernalia automatically constitutes the *corpus delecti,* and from there the prosecution may proceed to show, through confessions or other evidence, who the perpetrators of such crimes were.

Many security authorities feel that a certain percentage of employee crimes should be prosecuted, as prosecution has a therapeutic effect on the rest of the work force. Failure to prosecute simply means that the severest penalty to be encountered is loss of one's job. This is a small price to pay for the professional industrial thief who goes from job to job in order to rip off his employers and, in effect, make the majority of his income from a life of crime.

SECTION IV

MAKING THE COMPANY WHOLE

Chapter 11

Recovery Procedures

Having accomplished his primary objective of separating thieves from the payroll, the security director should turn his attention toward attempts to make recoveries on behalf of the company, and also toward instituting certain procedures and controls which will act as a future deterrent to a repetition of the dishonesty. New security procedures and controls are more easily accepted by management following a major investigation than at any other time. The security manager can and should follow the old adage of "striking while the iron is hot."

On the other hand, very often recovery attempts will become quite involved and almost as time-consuming as certain phases of the original investigation. Many thieves will readily accept their discharge from the company and even bow to prosecution by a plea of guilty, but will rebel when pressure is brought to make a substantial financial recovery which could easily involve all of the offender's assets.

RESTITUTION

In many major investigations, a management decision to keep some of the lesser offenders on the payroll may be a prudent one. Often, to separate all of the offenders from the company would mean a crippling of operations and would be entirely counterproductive. For those offenders who are retained on the payroll, it must be understood by all parties that restitution in a mutually agreeable amount is imperative. Many companies also require an offender who is ticketed for rehabilitation to sign a consent to periodic polygraph tests (see Figure 10-2 in previous chapter).

Negotiations

On the question of restitution, it must be kept in mind that in most theft cases, the amounts of theft recorded, even in the best of confessions, are often estimates. For this reason, the exact amount of restitution should be negotiable between the director of security and the offender. The personnel department of the company should normally sit in on such negotiations, but the actual negotiations should be conducted by the security chief and no one else. To turn the matter of negotiated restitution over to the personnel department, or to some third person who was never a party to the investigation or the confession, is ill advised. It will only result in an unrealistic amount being demanded or agreed upon.

The main thrust of any negotiation regarding restitution is to insure that there is complete understanding among all parties that restitution is not being asked or given as a trade-off to criminal prosecution. The company should make clear that any restitution is clearly in recognition of the offending employee's civil obligation and not any criminal liability that he may possess.

As a practical matter, however, most prosecutors would never entertain the thought of prosecuting an employee once the company had entered into a restitution agreement. The only question of immunity at this point is that of immunity against loss of the offender's job. A suggested form for restitution from retained offenders is shown in Figure 11-1.

In the case of persons who are prosecuted, occasionally overtures will be made to the company by the offender or a close relative to accept restitution prior to the case coming to

CONSENT TO RESTITUTION

I, ________________________________, confirm the previous statement (s) dated ____________________, acknowledging thefts of property of____________ ____________________________, and to the best of my knowledge, the total amount of____________________ property that I have stolen is $____________. I further confirm my wish to reimburse______________________________, for the value of the property that I have stolen, and I hereby promise to pay the sum of $________________ @ the rate of__________________________. I further authorize ____________________________, to withhold such sum from my salary, applying the amount so withheld in reduction of this obligation.

It is understood that this acknowledgement of my obligation and promise to repay, is made solely in recognition of my civil obligation to________________ ________________________,and that no promises of immunity from prosecution have been made to me by______________________________, or its officers or agents.

Date:____________________ Signed:______________________________

Witnessed:

____

____

Figure 11-1. Suggested form for restitution from offenders who are retained as employees.

trial. If such an overture is received, it should be reported immediately to the prosecutor, and no acceptance of the offer should be made without the prosecutor's full blessing. Because of future relations with the police and prosecuting attorney, it would be a grave mistake for a company to accept money from an offender in a unilateral action.

If money is to be accepted at all under these circumstances, it should be accepted in whole and never in part. The company should take the position that it is not a collection agency and should demand payment in full, leaving the burden of financing the obligation up to the offender.

The Promissory Note

Occasionally, in isolated cases, the security director will realize that a significant and substantial amount of money can be acquired through the recovery process prior to even making contact with the prosecutor's office. In these cases, it is suggested that a promissory note, such as is shown in Figure 11-2, be signed by the offender. In utilizing such a note, the security official should insert a rate of interest in order that the note be legally binding. The company should further take the position that payment of the note must be made within several days, and in full.

Here again, the problem of raising the money for financing of the obligation is left entirely to the offender and/or his relatives. If a cash settlement is offered by a relative, the company should not accept it as such. Instead, the company should demand that the money be advanced to the offender and payment of the funds made directly by the offender to the company.

As with a restitution agreement, the vast majority of prosecuting attorneys would never accept a case in which a promissory note had been rendered and payment made. The security director should keep in mind, however, that in many states it is mandatory to report a felony which has been committed. Although this law is seldom enforced, especially in the context of industrial theft losses, the security chief should be

NOTE

$.. Date.., ...

On demand, for value received, I ....

promise to pay to ...., *or order,*

the sum of .... *Dollars*

in lawful money of the United States, with interest at the rate of .... *per cent per annum, interest payable monthly.*

In case said interest is not paid when due, it shall be added to the principal and bear a like rate of interest until paid and the whole of said principal and interest shall forthwith become due at the option of the holder hereof without notice.

All sums shall be paid at ....

The consideration for this note is money owed by me to ....
and is not given in exchange for any promise, express or implied, to withhold or stifle criminal prosecution.

In case suit is brought to collect this note or any portion thereof I promise and agree to pay such additional sum as the court may adjudge reasonable for attorneys' fees in said action.

....

....

....

Witness.. *Address*..

Figure 11-2. Promissory note to be signed by offender.

aware that technically he could be compounding a felony by accepting restitution and failing to report the felony to proper authorities.

As stated above, the company should demand payment in full. It is an old tactic for an offender to come in, for example, with an offer of one thousand dollars cash on a five thousand dollar obligation. Psychologically, he is banking on the fact that the security chief will accept the thousand dollars in hand. The offender realizes that once any part of the restitution has been accepted, he has very little to worry about, practically speaking, in the way of prosecution. It then becomes a struggle for the company to effect collection of the other four thousand dollars.

Another good use to which the promissory note can be put is in the presentation of a bonding claim, which will be discussed later in this chapter.

Restitution

Where the security chief feels that the offender has assets which could be converted in order to liquidate the obligation, or where he anticipates problems in effecting a collection, he may wish to take his chances on a strong prosecution in the criminal justice procedure.

In the case of first offenders and many second-time offenders, the courts will often order restitution as a condition of probation. The security director has to be extremely cautious so as not to make it appear that he is simply using the prosecutor's office and the courts as a collection agency. In other words, he should take the position that his primary interest is in furthering justice, and that any restitution which may be obtained through the probation is only secondary. As a practical matter, however, the entire criminal justice procedure can often work to the benefit of the injured company.

Civil Suits

After the smoke has cleared from the battlefield of discharge and prosecutions, the company may wish to entertain the thought of recovery through civil court action. Many states have come to recognize the problem of industrial theft, and a few have passed statutes for up to treble damages in such instances. Civil suits are easily undertaken and involve little or no risk to the company. They are best handled by a lawyer who specializes in collections. Such an attorney is a much better choice than ordinary corporate counsel, as he will normally work on a percentage of the recovery. Often the only costs to the company are filing fees.

Seldom do the offenders force such issues to go into full-blown court proceedings. Experience has shown that most of these actions result in default judgments, which then become routine collection matters for the company's attorney. Here again, of course, the existence of a promissory note or good formal statement enhances the attorney's negotiating position. In actions of this type, the recovery efforts are usually stretched out over many years, but they are handled by the attorney and require little or no attention from corporate officials.

THE BONDING CLAIM

Potential Problems

Of all the different recovery options open to a corporation, the one which can sometimes prove the most troublesome is the bonding claim. The settlement of a bonding claim can produce much misunderstanding and ill feeling between the company and its insurance carrier, simply because the company does not have a clear understanding of the insurance carrier's role when such a claim is presented.

In regular crime coverage by an insurance company, such as burglary and robbery insurance, the carrier is insuring the company against the stated risks. When an incident occurs, it is usually a matter of official record and there is little or no question regarding the happening. Claims of this type usually are settled promptly, without question, and with residual good feelings on the part of both participants.

In fidelity insurance, the role of the bonding company becomes somewhat different. Although the fidelity bond itself is bought and paid for by the company, from the point of view of the insurance carrier, the concept is that the carrier is, in effect, insuring the employees of the company against defalcations, rather than that it is insuring the company against an outside threat.

Most insurance carriers in the past ten years have found the fidelity bond business to be generally unprofitable and refuse to write such insurance, unless they feel compelled to do so in order to retain other profitable insurance coverage purchased by the company. In defense of the insurance carriers, it must be pointed out that there have been instances where companies have attempted to use fidelity bond claims to recoup losses suffered through mysterious disappearance, general inventory shortages, and the like. Some security persons have also been guilty of unscrupulous and unethical practices in securing inflated confessions from employees, only to attempt to make money for the company at the insurance carrier's expense.

Normally the insurance carrier is not a party to the closing of a major investigation, nor should it be. It must be kept in mind that it would be in the insurance carrier's interest to hold down the amounts of the confessions, whereas, of course, it would be in the interest of the security chief to develop the confessions to their fullest extent.

In other words, the bonding claim is the end product of efforts expended by the company, without any collaboration or consultation with the insurance carrier, to obtain a cash settlement from the carrier under the terms of the bond. For this and other reasons stated above, most insurance carriers are initially cool and unhelpful when such a claim is received. They tend to enter negotiations on such a claim from an adversary position rather than that of a cooperative partnership with the company in resolving the problem at hand.

Factors Influencing Settlement

Few, if any, bonding claims have ever been settled on one hundred per cent of face value; rather, they are settled on a percentage of the total claim. Such percentage is usually determined through many negotiating sessions, and is influenced to a great extent by the strength of the security director's case. In other words, the stronger the claim, the higher the settlement.

Inventory Records. Most fidelity insurance policies are clear in pointing out that no bonding claim will be accepted on the basis of a shortage in inventory records. On the other hand, once the claim has been made and accepted by the bonding company, the security director can utilize inventory records to bolster his case. Conversely, a bonding company reviewing and negotiating such a claim will almost always ask for inventory records, and if these are not supportive of the claim, the net result is a weakening of the company's overall position.

Stated another way, inventory records cannot be utilized as the basis for a claim, but will ultimately serve either to enhance or detract from the security director's negotiating position.

Production Records. Although almost all fidelity policies have the standard prohibition against utilizing inventory records as a basis for a claim, many are silent when it comes to production records, and it may be that a company's production records could very well become the basis of a bonding claim.

Accounting Records. Accounting records, unlike inventory or production records, are generally irrefutable. In a bonding claim based on an embezzlement, the accounting records can serve as the entire basis of the claim itself, even without a confession.

Undercover Investigation. Most astute claim adjusters will raise the question of undercover efforts early in the negotiating process. In other words, they will wish to obtain the entire story of the investigation, and if undercover was utilized, they will demand copies of the undercover reports. The security director has little choice in this matter; refusal to release the reports only weakens his negotiating position.

On the other hand, a favorite tactic of most bonding companies is to take the position, based on the undercover reports, that any thefts which occurred after the initial discoveries by the undercover agent cannot be included in the claim. This is based on the standard clause in most bonding contracts that once a theft is discovered by the employer, the employee is no longer bonded, and the insurance company is no longer financially liable for any acts which result from that time on.

There are two negotiating positions which the security chief can take in rebuttal to this posture by the insurance company: 1) that the undercover agent's reports were not available to management until the conclusion of the investigation; 2) that it was only reasonable for the company to wish to identify all of the thieves rather than precipitate a premature exposure of only one or two.

If the amount of thefts that occurred after the undercover agent's initial efforts is significant, then the company may have to resort to pressure, through its insurance broker, on a higher level of management within the insurance carrier. This can often be accomplished if the insurance carrier is providing other forms of insurance for the company and its relationship with the company can be viewed in an overall context rather than just on the basis of fidelity bonding coverage.

Here again, as in the case of inventory records, undercover reports can serve to bolster the claim but at the same time can often reduce the size of the claim if the carrier is successful in eliminating a significant number of thefts from the total.

Arbitration and Court Findings. A successful labor arbitration ruling or a guilty finding by a criminal court will greatly bolster the bonding claim. On the other hand, a not-guilty finding by a criminal court will weaken the claim. An adverse ruling by a labor arbitrator simply makes the bonding claim a moot question, as the offender is usually back on the payroll at that point.

Validity of Confession. With the possible exception of some embezzlement cases, the central core of any bonding claim will be the confession and how much weight should be attached to its validity. All of the records and reports involved in the foregoing discussion generally tend to support or strengthen the weight of the confession, but, on the other hand, if sufficient thought is not given to them beforehand, they sometimes can be used to subtract from the strength of the confession in supporting a claim.

Before starting negotiations with the insurance carrier, management should realize that the negotiating tactics of the claims adjuster will be to cast doubt on the validity of the confession, especially on the amounts of merchandise and/or money which are specified in such a statement. The more doubt that can be cast in this direction, the less money will be paid by the carrier to the company.

As a rule of thumb, a good handwritten statement by an offender, with little or no supporting evidence, is worth fifty cents on the dollar at best. In many cases it will be worth less than that. On the other hand, a good typewritten formal statement (question-and-answer style) with an impressive chain of supporting facts such as undercover reports, successful prosecution, inventory records, etc., is worth somewhere around seventy-five to eighty cents on the dollar. If management realizes this before even starting negotiations, their overall attitude and relationship with the insurance carrier will be on a much more amicable plane when the negotiations are concluded.

Deductibles

In virtually all fidelity bond policies there is a deductible clause. In computing the mathematics of the settlement, it is imperative that the deductible be applied to the total claim prior to any discounting of confessions. In all probability, the total claim will be based on amounts stated in written confessions by the offending employees. If these confessions are discounted, as discussed above, before the deductible is applied, the deductible will be reducing a lesser claim than the original claim. Therefore, it is in the interests of the company to have the deductible applied initially and resist any attempts by the adjuster to apply the deductible after a discounted amount has been agreed upon in relation to the confessions.

Most policies provide that the deductible is based on a per person involvement except where there is a collusive effort. In other words, where five employees have stolen separately and independently of each other, five deductibles would be applied to the total claim. On the other hand, if four of the five acted in concert and in collusion with each other, then only two deductibles would be applied; if all five were involved in a common theft ring, then only one deductible would be applied.

This points up the desirability and the necessity for developing complete information in any corroborative statements by co-offenders. This is one of the reasons why co-offenders should be present during the taking of formal statements and should sign as witnesses to each other's signatures on their respective statements. It becomes increasingly difficult for a bonding company to attempt to discount a confession when four other offenders have made essentially the same confession and corroborated in almost every detail each other's involvement.

Bonding Company's Recovery Attempts

After the settlement of the claim, the bonding companies almost always attempt recovery efforts of their own. This is normally referred to as "salvage." Under the terms of most bonding contracts, the bonding company is obligated to attempt to make a recovery on behalf of the employer, but only after satisfying their own reimbursement for moneys paid out in settlement of the bonding claim. In other words, the bonding company must be paid first, and if there is anything left over it goes to the employer. By the same token, if the security director has been successful in obtaining even partial restitution from one or more of the offenders who are included in the bonding claim, such recovery must be shown as a credit in computing the final settlement of the claim.

Premium Increases

It is the practice of most insurance carriers who write fidelity bond insurance to raise the premiums or the deductibles, or both, following the payment of claims. When the time for renewal of a policy arrives, the insurance company will review its experience with the particular company and, based on such experience, will raise the premium and/or deductible, or possibly leave it unchanged if no claims have been submitted. Knowing the realities of the situation, many company financial officers have taken the position that bonding insurance should only be utilized for a true catastrophe, and that routine bonding claims based on the efforts of the security department should not be put forward to the insurance company.

However, if a company has never submitted a bonding claim in the past and has continually paid bonding insurance premiums over a long period of years, there is no question but that the bonding company has made a substantial amount of money in writing the bond. Based on the premise that an efficient and progressive security department can ultimately clean up most thievery in a corporation, it is the author's opinion that most bonding claim issues should be submitted, even though they precipitate increased deductibles and/or premiums. Eventually the deductibles will reach a level where the insurance truly becomes catastrophe insurance, but by then most of the routine thievery should have been cleaned up in the company and this high deductible would have little impact.

Prolonged Negotiations

A final note that should be borne in mind by corporate officials negotiating for settlement of a bonding claim is that when a claim is first presented, the bonding company will automatically set up a financial reserve for a realistic maximum on the case. From that point on, it is in the interests of the bonding company to drag out negotiations on the settlement for as long as possible, because, in effect, they are operating on money that has already been earmarked for the corporation in question rather than for themselves and, of course, which is interest-free. By improving their cash flow in this manner, the insurance carrier can thus reduce short-term borrowings at the prevailing prime rate. It is believed that this policy is followed by the majority of major fidelity bond underwriters, and is one of a number of reasons why it becomes difficult to effect a speedy settlement of a bonding claim. Of course, if the final settlement reached is below the maximum amount of the reserve which had been set up, the difference reverts back to the insurance carrier as an added profit.

In conclusion, it is clear that a major responsibility of a security director is to effect financial recoveries for his company whenever possible. Obviously, he has a number of options, but it is just as obvious that he must be well versed in the various strategies and legalities that are involved in any of these processes.

Chapter 12

Post-Operative Therapy

Having made recovery efforts for his company, the security director should not consider his task complete until he has laid the groundwork and instituted programs for the future; programs which will be designed to reduce the chance of a repetition of the "illness" which his company has just endured.

Many corporate security departments are oriented toward investigative procedures and little else. Over the years, they have developed excellent expertise in investigative technique and have established outstanding track records for the apprehension of dishonest employees. However, their task is a never-ending one as they simply go from one major investigation to another. On the other hand, other corporate security departments have been noted for their strong approach to preventive and control procedures, with little emphasis and almost no expertise in the field of investigation. A disproportionate reliance on either approach, by itself, does not work in the best interests of the corporation.

Enlightened security management, and even corporate management as a whole, will recognize and accept the fact that good selection procedures and internal controls are basically designed for honest people—those in the center of the spectrum whose propensities toward dishonesty or honesty can go either way. This same enlightened thinking will also accept the fact that if a company has confirmed, hard-core thieves on its payroll, normal security controls and procedures will not deter the hard-core element from thievery unless the company is prepared to operate a penitentiary. The controls will, ironically, cause the thieves to become even more adept in their efforts and more difficult to apprehend.

In other words, controls are only as good as the people who administer them, and when the people factor is taken into consideration, one immediately must recognize the existence of human weaknesses and human error. Thieves recognize this, and wait for the opportunity to take advantage of any weakness or loophole. Therefore, the correct approach to the whole problem of industrial and business security is a blending of the two—investigation and controls. The correct proportions of the blend will be determined largely by the nature of the company's business and the make-up of its work force.

APPLICANT SCREENING AND EMPLOYEE SELECTION

It is absolutely essential in building a healthy company to begin with the selection of persons who are inherently honest, or at least basically honest if the proper controls and procedures are in effect. Hiring persons who have a history of consistent employee theft only insures that security problems will continue to develop and that there will always be plenty of work to occupy the investigation section of the security department.

Reference Check Limitations

Much has been written about the desirability of obtaining sufficient references on prospective employees. The plain truth of the matter is that seldom do such reference checks reveal any history of dishonesty. One reason is that most employee dishonesty goes undiscovered and is simply absorbed into the company's business; it is ultimately passed along to the consumer in the form of higher prices. Another reason is that, unfortunately, most employers are reluctant to release derogatory information about former employees out of fear of lawsuits. Consequently employers have come to put little faith in the traditional reference letters because they themselves refuse to be candid in responding when such questionnaires reach their own desks.

In addition, America has experienced great shifts in population since World War II; ours has become a mobile society. Job applicants no longer generally come from an employer's immediate community; thus the employer must seek some method of screening prospective employees.

Most police departments and the FBI do not make their criminal files available to the private employer. This has come about because of the federal Privacy Act, as well as administrative prohibitions on police departments imposed by the Law Enforcement Assistance Administration. Many companies have found that the background investigation conducted by private agencies often reveals little that is not already known to the employer. Such investigations are particularly susceptible to influence by malicious neighbors, vindictive past employers or disgruntled former fellow workers. Clearly, such a procedure is inadequate from the employer's view; equally important, it is fraught with the danger of injustice to the job applicant.

As is discussed in Chapter 7, quality background investigations can be very meaningful if the company is prepared to pay for such work. Such an investigation is conducted through field work and personal interviews. It is not conducted by telephone, and its cost may well exceed $200.

Polygraph as Screening Tool

As a result of these hard realities, many employers have turned to the polygraph to screen job applicants and to aid in internal investigations. Most of these companies restrict polygraph use to sensitive positions, i.e., those employees who will have access to money or merchandise. Partly because of the cost of a pre-employment test (from $25 to $50 nationally in 1978), polygraph is not used or even advocated for the great masses of job applicants. It is principally beneficial to companies which are particularly susceptible to internal theft. It is interesting that the use of the polygraph to screen job applicants over the past twenty years has increased in direct proportion to the rise of employee theft.

There are tens of thousands of companies today that utilize polygraph screening for job applicants. There are probably hundreds of companies that have attempted polygraph screening but have dropped the program because of poor experiences. In looking at these successes and the failures, one can only conclude that the successes were attributable to a sound, ethical program whose propriety could not be successfully challenged. The failures, on the other hand, can be attributed to a number of factors such as poorly qualified polygraphists, improper administration of the program by the company itself, over-selling of the program initially, successful challenges by organized labor and poor quality control within the polygraph agency.

Acceptance of Programs. That a polygraph program can be administered successfully is attested to by the experience of one large nationwide corporation with sales approximating three billion dollars per year, the McKesson companies of Foremost-McKesson, Inc., San Francisco. This company started its nationwide polygraph program in 1954 and averaged two to three thousand job applicant examinations per year in the intervening time. The program is still going strong. In one major division of the corporation, a number of senior management personnel who were screened by the polygraph technique when they were job applicants are now among the biggest boosters of the program.

There are other smaller companies which have pre-employment polygraph programs dating back to the 1940's. Here again, the programs are still in effect and going strong. On the whole, employees themselves do not resent a well-administered, ethical pre-employment screening program. In survey after survey of persons who had just undergone examinations for a job, the lowest positive rating ever obtained was approximately seventy-five per cent in favor of what the company was attempting to do through the use of the polygraph.[17] Certain politicians and civil libertarians who would choose to outlaw the use of the polygraph in business and industry have consistently chosen to ignore the fact that American workers themselves overwhelmingly give it a positive rating. These polygraph opponents have decided what is best for the public, even though it is contrary to the opinions of those people most affected.

If the security department has sufficient expertise in the polygraph field, then a polygraph program should probably be administered and controlled by the senior polygraph professional on the corporate staff. Otherwise, the program can be administered and monitored by either the security department or the personnel department. At any rate, the administration of an ethical and proper program involves constant monitoring of the entire procedure.

Proper Job-Related Questions. The average person applying for a job which requires a polygraph screening test can usually appreciate the importance and necessity of job-related questions. Generally, people tend to realize that this much of their privacy must be surrendered in view of the position of trust for which they have applied. On the other hand, these same people can be quick to sense when the polygraphist is straying into areas which are not so clearly related to the job at stake. This, then, is one of the most important areas of which the polygraphist must be cognizant in his relationship with the examinee.

One of the most basic problems in the polygraph field is the occasional tendency of the polygraphist to want to impress his client with his apparent ability to develop all sorts of derogatory information concerning the person being tested. This

---

[17] J. Kirk Barefoot (Ed.), *The Polygraph Story*. Linthicum Heights, Md.: American Polygraph Association, 1974. (Survey by Chicago Professional Polygraph Center, 1973.)

desire often manifests itself even though the applicant is ultimately recommended for the position. For instance, one occasionally sees polygraph reports on job applicants in their twenties where the only bit of derogatory information is, "at the age of nine, subject shoplifted some candy from a local store," or "subject admitted that at the age of thirteen she stole $4 from her mother's purse."

From a professional viewpoint, there is absolutely no reason for any polygraphist to delve into the distant past history of an applicant in that age bracket and to that extent, especially where there is no history of stealing in the immediate past, and where the results of the polygraph examination are negative (indicating no deception). Such practices on the part of the polygraphist only leave him open to speculation as to the real reasons for exploring a person's background to that depth.

It would be much better for the polygraphist to assess the person's work background in relation to his age, and to frame questions covering a more reasonable period of time. For example, if an examinee is in his or her mid-twenties and has held jobs during the past four or five years where temptation was present, the examiner would be inclined to frame his questions within that time period. He should not have any interest in delving into thefts or other irregularities prior to the age of eighteen.

All polygraphists should keep in mind that, as a general rule, people's moral values and attitudes toward dishonesty are in a state of flux during their teens and only tend to become firmly established toward the age of eighteen and beyond. To ask a thirty-five-year-old applicant for an executive position about irregularities committed during his teenage years would not only be completely irrelevant, but would be degrading and repugnant to the examinee. Regardless of the outcome that examinee would never become an enthusiastic supporter of pre-employment polygraph testing. Most adults prefer to be judged on their record and not on their childhood; all polygraphists should keep this in mind.

On the other hand, it is more difficult to establish the risk factor of a person eighteen or nineteen years of age. Here, of necessity, the polygraph examination must delve back at least several years to the ages of sixteen and seventeen. If such a person had committed some irregularities at the ages of fourteen and fifteen but then had refrained from such behavior during the ages of sixteen and seventeen (while still having the opportunity), the inclination would be to base a judgment on the past two years. By the same token, the examiner should not disclose in a written report to this client the derogatory behavior noted between the ages of fourteen and fifteen. After all, this information is almost never available from police departments, and polygraphists should not attempt to put themselves in a special position of dredging up and making available to clients a complete juvenile history of a job applicant.

Some polygraphists, because of their personal avocation of law enforcement or their own religious background, are inclined to allow their judgments to be influenced in making recommendations to their clients. They permit their judgments to be based on their own personal standards rather than the realities of the job marketplace. A person's extramarital affairs or sexual adventures are certainly no business of the polygraphist in private practice, and an inquiry into sexual habits or sexual deviations has absolutely no place in the private employment sector.

Questions concerning arrests and convictions, as well as drug usage, should have some reasonable bearing on the job for which the examinee is applying. Using the example of the thirty-five-year-old executive who has been able to establish to the examiner's satisfaction that he has been relatively honest in all of his jobs for the past ten years, there is absolutely no sense in asking, as a general question, whether or not he has been arrested. It would be much more proper to phrase such a question, "In the past fifteen years, have you been arrested for any offenses other than traffic violations?"

In some jurisdictions, arrest questions are now prohibited and the polygraphist must confine himself to questions on convictions, and then only questions that have some bearing on the job at hand. Again, questions on employment history should have some reasonable time limitation in relation to the applicant's age.

A few polygraphists occasionally get carried away with their talents at diagnosing physical and mental problems in the examinee. Questions about physical and mental conditions, again, must be job-related if used at all, and then should only be directed at omissions that the examinee may have made in filling out a medical questionnaire.

Explaining Non-Hire Decisions. Communication with applicants regarding hire/non-hire decisions must be handled carefully. Personnel people must be trained to advise all job applicants that a number of people are being considered for the position and that a number of factors will be taken into consideration before any determination is made. If a discussion of the polygraph comes up, the personnel director should point out that the security screening process is only one of a number of factors upon which the company will base its decision.

Following the pre-employment test, the personnel director or his representative should *never* advise the job applicant that he or she will not be hired because of a failure to pass the polygraph test or even a failure to meet the security standards. The personnel director is under absolutely no obligation to advise the applicant as to a specific reason for not hiring. As long as it can be conclusively shown that the refusal to hire is not based on race, religion, ethnic background, age or sex, no specific reason need be given. There have been all too many incidents over the years of a personnel department advising the job applicant that he could not be hired because of the results of his polygraph test. This has often resulted in the job applicant returning to the polygraph laboratory with a view toward venting his anger and frustration upon the polygraphist.

In the case of one large company where the pre-employment examination is administered after hiring but during the probationary period, the standard explanation is: "You have been here on a trial basis and we have decided at this time that you are not suited to our type of business."

In conclusion, there is no other form of background investigative procedure that can match the effectiveness of a well-administered pre-employment polygraph program.

Written Honesty Tests

In those jurisdictions where pre-employment polygraphing is not possible due to legal restrictions, or even in remote areas where polygraph services are not available, a company should seriously consider the use of a paper-and-pencil honesty test on job applicants. As in polygraph testing, written honesty tests should normally be restricted to those persons having access to money or merchandise.

There are several good tests on the market, such as the Reid Report.[18] Most authorities in the field believe that there is no equal to a good polygraph pre-employment test, but that a paper-and-pencil honesty attitude test, such as the Reid Report, is the best possible substitute. On a dollar-for-dollar basis, there is no question but that the written honesty test is a viable alternative if polygraph is illegal or unavailable.

INTERNAL CONTROLS

Internal controls almost invariably come about because of the dishonest activities of company employees. Probably ninety per cent of all internal controls in existence today in any company could be traced to an incident of dishonesty in that company or in some other firm. In other words, if all employees were inherently honest, there would be no controls and companies would never give a second thought to internal security measures.

Within the scope of this text, it would be impossible to detail the type of controls which should be set up for any particular company. Controls will vary from corporation to corporation and will be based largely on the company's security history, lessons learned from other corporations, and the nature of the company's business.

---

[18]See footnote 4, Introduction, page 8

While the amount of the "paperwork" portion of internal controls will have to be in balance with efficiency and cost, internal controls act as a strong deterrent to employee theft. Not only is there a psychological barrier raised, but the "paperwork" also helps provide an audit trail for company investigators. To keep internal controls at an effective level, they should be tested frequently by management or security personnel; e.g., "extra" merchandise added to a driver's load; "special" shipments into the company's own receiving department, where no purchase order is being held; an executive's attempt to circumvent normal house order procedures to test employee awareness. Finally, management should conduct an ongoing routine program of investigation of shortages or other irregularities. Such a program mentally reinforces the concept of controls in the employees' minds and creates the most important realization of all—management *does* care.

Security Inspection Check Lists

To give the reader an idea of the type of monitoring that should be done during a security inspection or a survey, a Security Inspection Check List for apparel manufacturing locations is shown in Appendix 1 at the end of this chapter. A similar form for use in a retail environment is also presented in Appendix 2. Many security managers will be quick to realize that additional points could be added to either of the forms presented. However, it must be remembered that a different type of company would have different needs.

The important thing to note is that the questions are designed to get the inspector thinking in depth about the various logical security areas of the company's operations. Additionally, if the inspector will attempt to think beyond each question itself, and consider the question's *intent,* he will be able to formulate additional questions of his own. In this way, he may uncover weaknesses which would not be spotlighted by a person merely adhering to the printed format.

The reader will also note that both check lists start with questions on the personnel function. If the security department has been given sufficient latitude, certain personnel procedures should be included for the simple reason that good security begins with the selection of job applicants who are basically honest.

A well-done security inspection provides one additional dividend for the security manager—a future intelligence reference for either overt or covert investigators.

Employee Education and Motivation

One nationally known security consultant has consistently maintained that company security departments should attempt to win over at least one-third of the employees in the battle against employee dishonesty. He has further stated that if as many as fifty per cent of the employee work force can be converted to the side of security and honesty, then the war against employee theft has been won.

Opinions of other experts on these percentages and their ultimate effects may very well vary, but generally all are in agreement that a certain number of the employees must be won over in order to make the security department's job meaningful. This can be done in a variety of ways, such as posters, pay envelope inserts, reward programs, and face-to-face meetings.

Experience has shown that it is generally easy for security department personnel to address employee groups on the question of dishonesty and its full ramifications, including its effects on the employee's family. It must be pointed out, however, that there is usually a reluctance on the part of regular management to discuss matters of honesty and dishonesty with the rank-and-file employees. This is nothing more than a morale problem on the part of management people. They should be helped to understand that when it comes to matters of stealing, breaking the rules, breaking the law on company property, and other security violations, the employees themselves are quick to discuss these things at every given opportunity. Realizing this, security personnel have found it easy to have face-to-face talks with employee groups. For the same reason, it should be easy for management to do likewise. The need for education thus applies to management as well as regular employees.

In locations where the employee work force is so great in numbers as to make such meetings impossible, monthly safety/security meetings can be instituted. In many companies utilizing the safety/security committee concept, members of the committee representing the rank-and-file employees are rotated every two or three months so that eventually a good cross-section of the employee work force has had some contact, either direct or indirect, with the security committee.

Employee education should be viewed as a never-ending process and one that is necessary in order to ultimately develop the degree of motivation that is desirable with the employee work force.

The Role of Management in Cooperation

No security program can ever be effective unless it has the complete backing of top management personnel. There probably always will be some members of middle management, as well as certain segments of the employee work force, who will automatically be anti-security. For this reason, it is imperative that any security program which is adopted has the backing of the most senior members of management in the corporation.

Many people in management today regard it as somewhat immoral if temptation is constantly placed in front of employees without necessary restraints and controls accompanying it. They further believe that the company has an obligation to its honest employees to help them remain honest, and that to turn away from this obligation is to be unfaithful to their employees as well as their own moral principles as professional managers. In some companies, senior management personnel have adopted an even more pragmatic viewpoint toward security, either in addition to the above or in place of it—namely that the security department should be considered in the same light as any other profit center and that it has the capability of putting profits back into the bottom line of the company's profit-and-loss statement that would otherwise be lost.

There is no question that the private security industry, as a whole, has as much if not more potential to affect the economy of the country and the entire cost of living on a day-to-day basis than the remainder of the whole criminal justice system.

Within management and the private security industry there must be a recognition that, despite good planning procedures, controls and screening, there will always be a small element of the work force dedicated to theft or "ripping off the boss." Because of this, there will always be a need for good investigating capability. The modern day private enterprise system is too complex to think otherwise. Employee or internal theft must be investigated and exposed in order that overall progress in the war against internal crime be continued.

APPENDIXES

Appendix 1

SECURITY INSPECTION REPORT

Date of original inspection ______________
Date of last inspection ______________
Date of this inspection ______________

COMPLETE IN QUADRUPLICATE:
ORIGINAL—Plant Manager
DUPLICATE—
TRIPLICATE—Corp. Security
QUADRUPLICATE—Regional Security

Name & Address
of Location

____

Officials interviewed during survey

(Names)	(Titles)
______________________	______________________
______________________	______________________
______________________	______________________

Instructions: All "NO" answers to be explained on separate page. Also, all descriptions called for and any question where improved control is needed.

Petty Cash Count:
Cash authorized ______________________
Cash on hand ______________________
Receipts & Vouchers ______________________
Over/Short ______________________
Volume in last fiscal year ______________________

$ Overage or (Loss) in last fiscal year	Raw Materials ______________
	In Process ______________
	Finished Stock ______________

Total Number of Personnel (everyone headquartered in building)

Warehse	Factory	Drivers	Off.	Salrd Supvrs. & Exec.

Type of Building (describe fully)

Y N

PERSONNEL

1. Is the application form completed properly and signed?
2. Where a security screening program is in effect, have polygraph tests or Reid Reports been administered to all applications for sensitive positions?
3. Are references of applicants for sensitive positions checked by either letter or telephone?
4. Are completed reference letters or telephone checklists placed in the personal folder?
5. Are criminal activity checks made of applicants for sensitive positions?
7. Are there any company rules which have been put in writing?
8. Are they adequate for this location?
9. Are company rules posted?
10. Are disciplinary actions reduced to writing and placed in personnel files?
11. Are regular safety-security meetings held?
    Did you review the minutes of the last three meetings?
12. Is the Incident Report Form regularly in use?
13. Is the Monthly Security Inspection Report in use?
    Date of last report:
14. Has responsibility for overall security been delegated to one employee?
    Name:

GUARDS & WATCHMEN

21. Are guards or watchmen utilized at this location? Attach post No's and hours of coverage if any change since last report.
22. Is coverage adequate for this location?
24. Describe current rate of pay for guards and watchmen.
25. If agency guards what do we pay agency? ______________________
    What do the men receive?
27. Is there a guard manual of written rules and regulations for the night watchmen and guard force?

Y N

GUARDS & WATCHMEN (cont'd)

28. Are present instructions, rules and regulations adequate for current conditions?
(Attach any changes in guard manual since last report)

30. Is a daily log maintained giving full particulars of all persons entering and leaving the building, except at starting and quitting time?

31. Is the Guard's Daily Report in use?
Who receives the report, and takes follow-up action?
______
How long retained? ______

33. List type of firearms and other equipment carried by guards.
Is equipment owned by Company Guard Agency or personal property of guards?

36. Is instruction and training given to guards and watchmen in the use of firearms and emergency equipment?
By whom? ______

KEYS, ALARM AND WATCHCLOCK SYSTEMS

41. List key holders by name, position, locks or doors, type of key.
Number of master keys? ______
Where and how are spares secured ______
Normal opening and closing times of office door if not on alarm?

42. Are keys withheld from alarm company? If not is key sealed?

43. Are locks changed when a key is lost?

44. Are locks changed when a key holder leaves the company?
Date locks were last changed:

45. Are guard's building keys and gate keys maintained in secure place when not actually in use?

46. Are the buildings equipped with burglar or other type alarms? If so, describe type of alarms, i.e., central station and name of company, local alarm to police, F.D., outside bell, internal day alarms, etc.

47. List burglar alarm circuit No. area covered, title of card holders and normal opening and closing times by days.

48. Was a review of alarm company or time recorder lock reports of unusual openings and closings since last inspection found to be negative?

49. Are all alarm systems tested regularly? Did you personally make a test of all perimeter alarms?
Types of system tested? ______

50. Describe any changes since last report in floor-plan showing the following:
a) Entrances and exits (including overhead doors).
b) Exits connected with alarms and/or padlocked.
c) Watchman's clock stations, numbered in sequence.
d) Guard stations.
e) Fire stations.

Y N

KEYS, ALARM AND WATCHCLOCK SYSTEMS (cont'd)

51. Did you make a clock round with the guard?
Is the number of stations and location adequate?
Describe frequency of rounds, type system used, if central station controlled, etc.

52. How often are watch clock tapes or discs examined?
____
By whom?_______________
Are tapes or discs retained until they are examined by insurance inspectors and Corporate security officers?

TRANSPORTATION SECURITY

54. Are adequate security procedures taken when merchandise is left in trailers or box cars?

55. Is responsible person present on docks at all times when receiving and shipping doors are open?
Title:

56. Are shipping and receiving rooms properly separated?

57. At warehouse locations, are shipping activities, which are performed during other than normal working hours, subject to extra security measure?
Explain:

58. Are truck driver movements restricted from the stock areas?
(Describe amount of access).

59. Is a well defined and carefully executed plan in effect for the use of security seals on Company or leased trucks which are used to transfer products or finished goods between inter-companies, plants and warehouses? Describe plan, name person in charge of supervision of plan and method used for custody of seals.

60. If seals are not used, have paper controls been set up?
(If so, describe fully).

INTERNAL CONTROLS

61. Is a permanent access or pass system in effect for off hours?
Covers: (List by name and title)

62. Is employee identification adequate?

63. Is procedure for handling returns and/or seconds adequate for security purposes?

64. Does procedure for disposal of waste cartons, trash, etc. minimize opportunity for theft?
Did you inspect the trash?
(Describe controls set up)

65. If there are restricted areas, buildings, or offices on the premises, are these areas protected? (Describe)

66. Are janitors, maintenance men, and clean up crews properly supervised for control of theft activity?
(Describe)

Y N

INTERNAL CONTROLS (cont'd)

67. Are employees required to enter and leave the building only via the ________ door?

68. Are employees prohibited from taking lunch boxes, packages, purses or handbags, etc., into stock and working areas?

69. Are employees' packages which are brought into the building, stored in their lockers or other place approved by management?

70. Are the locker space and restrooms outside stock and working areas?

72. Is there local physical inspection of handbags, packages, lunch boxes, etc., made as employees leave the building?
Is the inspection complete as opposed to a spot check?
Did you personally observe or make an inspection?
Who makes the inspection?
____
Where is the inspection performed? ________________
How often is the inspection performed? ________________

74. Are employees prohibited from remaining in stock and working areas during lunch periods?

75. Is a supervisor on duty during lunch periods?

76. Is procedure for pick-up of employee purchases in conformity with good security?
(Describe fully)

78. Are employees' cars parked only in areas a secure distance from shipping and receiving doors?

80. Did you make an inspection of lockers, work stations, sensitive areas, etc.?
List names of questionable products, location and name of employee.

81. Are lavatories and refuse containers free of improper items?

82. Are all cases in the full case area full and unopened?

PERIMETER SECURITY

106. Did the perimeter security check reveal that proper security measures are in use? (Outside of buildings, ground floor, basement windows, flood-lights, skylights, doors, approach to roof, etc.)

107. Did you walk the fence line?
Is the present condition of fence perimeter adequate? List any holes, breaks, etc.
Describe height, type and material of fence.
____

108. When fence gates are not in use, are they kept closed?

109. When gates are not guarded, are they securely locked?

MISCELLANEOUS

115. List any security deficiencies remaining since last inspection.

116. List any personal observations during this Security Inspection that indicate

		Y	N
	MISCELLANEOUS (cont'd)		
	laxness in administration of security, gambling, drinking, leaving work place, sleeping on job, etc.		
117.	**Describe any unusual political or labor practices that could have some bearing on security.**		
118.	**On Security Department copies only, show date of your last personal contact with local police sources.**		

Submitted by: ______________________________

Appendix 2

SECURITY INSPECTION REPORT
Retail Stores Division

Distribution: (_________ copies)
Orig:
Dup:
Trip:
Quad:

Company: ____
Store Location: ____________________________________
Date: _________ **Prepared by** ____________________________
Sales: _________ **% Shrink** _________ **For Period** ____________________

Officials interviewed ____________________________________
during survey: ____
(Names) **(Titles)**

Instructions: **All "NO" answers are to be explained on separate page; include all descriptions required; note where improved control is needed. SN denotes See Notes; NA denotes Not Applicable at this location.**

Description of building: (Sq. ft. ____________________) ____________________
____
____

Number of employees in the building

	Exec	Office	Sales	Ship/Rec.	Alter	Wrap	Porters	Leased
Full Time								
Part Time								

Y N

1. PERSONNEL
   Does inspection of randomly selected personnel files verify that:
   a. The application form is properly completed?
   b. Adequate references have been checked?
   c. Where a security screening program is in effect, have polygraph tests or Reid reports been administered to all persons having access to money and/or merchandise?
   d. Fair credit notification, Policy acceptance, and Salary withholding agreements are signed?
   e. Serious infractions of Company rules are documented in writing by use of an employee consultation or disciplinary action form?
2. PHYSICAL SECURITY-ALARMS (external)
   This building is protected by ( ) circuits which transmit alarms to:
   a. Co.: __________
   City: __________ Phone: __________
   Cir# __________ covers __________

   Cir# __________ covers __________

   Cir# __________ covers __________

   b. Card Holders for:
   Cir# __________ are:

Name	Title

   Cir# __________ are __________

Y N

PERSONNEL (cont'd)

c. Cir#__________was tested and it transmitted an alarm as required. It took __________ minutes for (phone) (physical) response.
Cir#__________was tested and it transmitted an alarm as required. It took __________ minutes for (phone) (physical) response.

d. Operations should receive reports of *unusual* openings and closings. Did your review of the latter fail to reveal any pattern?

e. Have adequate security provisions been made for personnel required to come in early or stay late?

f. Describe any changes since last report in floor-plan showing the following:
Entrance and exits, incl. overhead doors
Exits connected with alarm.

3. PHYSICAL SECURITY—ALARMS (internal)

a. Do all non-public exit doors have audible panic type alarms?

b. Was each tested and found to be in working order?

c. Was sound level sufficient to alert management?

d. By-pass keys are possessed by:

Name	Title

4. PHYSICAL SECURITY—LOCKS

Locks must be changed when an identifiable key is lost or a key holder leaves the company. Padlocks should be kept in locked position.

a. Peripheral door keys are possessed by:

Name	Title

b. Were locks changed:
When the last key-holder left the company?
Whenever an identifiable key is lost?
Date locks were last changed ____________________

c. Detail lock repairs indicated:

Y N

5. DISASTER/THREATS EMERGENCIES
a. Has an organization for handling emergencies been established?
b. Are instructions adequate for this location?
c. Are supervisory personnel familiar with these instructions?
d. Does switchboard operator have appropriate instruction sheet and acknowledge understanding?
e. Date of last Fire Drill?
f. List any unreported incidents since last inspection:

6. EMPLOYEE CONTROL—ENTRANCE AND OWN/GOODS
Employees should be prohibited from taking own goods such as lunches, packages, outerwear and other personal belongings into stock, office and selling areas.
a. Is store free of above violations?
b. Are employees' cars parked only in areas a secure distance from shipping and receiving doors?
c. Are employees required to enter and leave the building only via the _________ door?
This should include lunch and rest periods.
d. Are employee lockers provided?
e. If so, is the location appropriate for their intended purpose?
f. Are lockers inspected?
By whom:
Frequency: Last time:
g. Where lockers are not provided, describe alternative security provisions for personal belongings.

h. Is there a physical inspection of packages, lunch boxes, and personal belongings, etc., made as employees leave the building?
i. Is the inspection complete as opposed to a spot check?
j. Who makes the inspection:
k. Where is the inspection performed?
l. How often made:

Y N

7. EMPLOYEE CONTROL—PURCHASES

a. Is procedure for employee purchases in conformity with good security?

b. Is a central holding area provided and use enforced?

c. What are the provisions to deter unauthorized discounts to friends and/or relatives: ______

8. OFFICE

a. Is the requirement to exclude unauthorized personnel enforced?

b. Is reserve stock of: cash refund; charge credit; merchandise due bills; own goods sales checks; and gift bonds adequately secured?

c. Safe combination holders are:

Name	Title

d. Date last changed:
Reason:

e. If alarmed, type

f. Is change drawer locked when not in use?

9. OPERATING FUND

a. Did inspection fail to reveal any merchandise or money IOU's or post-dated personal checks in the operating fund?

b. The daily cash report (dated)______ listing $______ corresponds with validated bank deposit receipt (dated)______ for $______

c. Describe the security provided for transporting money to bank.

d. Is manager provided with timely record of cash overages and shortages for each person or Register?

10. SALES RECORDING AND CASHIERING

a. Are you satisfied based on personal observation that sales personnel are adhering to established company sales recording policy?

b. Does this include the requirement that all sales are recorded prior to customer departure?

c. Does sales personnel compare signatures on all charged purchases with charge card or other I.D.?

d. Did your observation indicate an adherence to the requirement that all over-rings/void slips receive supervisor's signature at the time of occurrence?

e. Does night closing procedure include stripping cash registers of money, leaving drawers ajar, and an accounting for monies turned in?

f. Does the daily cash report list each register observed on the premises?

Y N

11. REFUNDS—Management should physically inspect all returned goods and affix his signature in front of customer.
    a. Is refund procedure for this location adequate? If not, recommend accordingly.
    b. Is sales ticket and/or cash register receipt defaced when credit voucher or refund is issued?

12. SHIPPING/RECEIVING
    a. Did your inspection of packages prepared for delivery (UPS, Parcel Post, etc.) contain only authorized and cash register validated labels?
    b. Where supplemental labels are required, are controls adequate to preclude unauthorized usage?
    c. Did your inspection of transfer procedures reveal adherence to established policy?
    d. Did spot check counts reveal discrepancies?
    e. Are transfer hangers, cages, and tanks sealable?
    f. Is there a well-defined and carefully executed plan in effect for the use of security seals on both transfer containers and company trucks?

(NOTE: Management should be reminded to inspect seals for tampering prior to removal.)

    g. Is the manifest prepared neatly, correctly and completely with attention to form and seal numbers?
    h. A representative of management should be present on the dock at all times when receiving doors are open. Except for truck loading/unloading, drivers should not be permitted in the receiving area. Are you satisfied these requirements are being met?
    i. When unresolved discrepancies in counts exist, is the policy to notify security, consistently followed?

13. WASTE DISPOSAL
    a. Does procedure for disposal of waste cartons, trash, etc., minimize opportunities for theft?
    b. Are boxes crushed at the department level?
    c. Does management periodically inspect trash?
    d. Does management personally supervise its removal outside?
    e. Results of your trash inspection:

14. DAMAGED MERCHANDISE
    a. Are existing procedures adequate to insure expeditious processing of damaged merchandise to minimize losses?

Y N

DAMAGED MERCHANDISE (cont'd)

b. Are controls sufficient to prevent employees from taking damaged merchandise for personal use?

15. SALES SUPPORTING AREAS

a. Fitting Rooms:

(1) Are unused rooms or banks effectively closed off?

(2) Describe any evidence of shoplifting:

(3) If improved controls are needed, describe:

b. Alteration Work Rooms:
Do all garments in the alteration department have an alteration tag plus sales check, own goods ticket or repair memo? List all discrepancies.

c. Display:

(1) Is the requirement to sign in and out major garments for display being enforced?

(2) Is all merchandise found in the display room properly documented?

d. Advertising:
Is the requirement to sign in and out major garments for advertising being enforced?

e. Forward Stock Rooms:
Holds—is tagging policy being followed which requires proper authorization, 3-day maximum and expeditious return to stock if not purchased?

16. SECURITY FILE
To insure compliance with corporate standards and minimize the opportunity for adverse litigation, each store should have a security file. Did your inspection of this file confirm that:

a. It contains a copy of the security manual?

b. The manager is completing his monthly Security Inspection Check List?

c. Incident Report Forms are being completed for each required occasion?

d. There were no unreported instances of suspected or actual dishonesty or known losses?

e. A supply of apprehension support forms are available?
These should include blank "confession" and "general releases."

f. It contains a list of emergency telephone numbers, including local police, fire and ambulance.

Y N

17. SHOPLIFTING

a. Are cash registers/check-out areas designed in a manner to preclude customer accessibility to cash drawers, bags, boxes and sales recording media?

b. Is valuable merchandise displayed with security in mind? If not, recommend accordingly?

c. Is arrangement of displays such as to preclude obstructed views?

d. If the security implications of this location suggest a need for the sealing of all customer packages, demonstrate to management how this can be accomplished in good taste by use of string handles taped to package.

e. What are the elements needed for a valid shoplifting apprehension at this location?

f. Have all members of Management been briefed in proper apprehension procedures and aware of the foregoing required elements?

g. Are you satisfied that there is a continuous educational effort directed at making employees aware of this problem?

h. Are new employees thoroughly familiar with the reward program?

i. Did they receive a personal copy of the reward brochure and anti-shoplifting pamphlet?

j. List other training aides being used:

k. If appropriate, recommend anti-shoplifting devices such as hanger guards, plastic ties, alternating hangers, observation ports, etc.

l. Is the anti-shoplifting coverage for this store adequate?

18. GENERAL

a. Are *Security Rules and Regulations* displayed prominently in this store?

b. Specific management request for additional security:

c. List any personal observations made during this security inspection that indicate laxness in administration of security, i.e., gambling, drinking, leaving work place, sleeping on job.

Y N

GENERAL (cont'd)

d. Is the store climate such to create proper spirit of discipline and alertness to the security problem?

e. What security problems discovered during previous visit still exist? ______

f. Describe any unusual political or labor practices which might have a bearing on security: ______

g. Local Police Department ______

(Phone)

(P. Larc. =) (G. Larc. =)

Age considered adult ______

(On security department copies only, show local police contacts and appropriate comments, if any)

NOTE:

The remaining portion of this Inspection Report pertains to Safety and compliance with the OSHA Act. When possible, deficiencies should be corrected "on-the-spot." If unable to do so, management must be advised of those deficiencies during your "Post Inspection" discussion with management.

19. HOUSEKEEPING & MAINTENANCE

a. Are all floors and stairways free of obstructions and slipping hazards?

b. Are all floor surfaces, carpets, tile, etc., in good condition, sidewalks adjacent to store free of slipping and tripping hazards?

c. Are stairs in good condition, handrails secure and lighting adequate?

d. Are all exit doors easily operated, exit signs and lights in good condition, exits unobstructed?

e. Is all glass, external and internal, including mirrors, free of cracks, chips, etc.?

f. Is all furniture in use in safe condition, solid, free of cracks, splinters, burrs, projections, etc.?

g. Are all ladders and fitting stands in sturdy, tight condition, free of cracks, splinters and defective hardware?

Y N

HOUSEKEEPING & MAINTENANCE (cont'd)

h. Are swing racks in good condition, easily operated?

i. Are utility closets orderly and free of debris?

j. Are toilet facilities bright, clean, ventilated, with adequate supplies?

20. ELECTRICAL SAFETY

a. Are electrical outlets ample so that multiple plugs and extension cords for other than temporary wiring are not used?

b. Are flexible cords in good condition and free of splices?

c. Are electrical cabinets kept closed and clear for 30" in front?

d. Has the electrical system been checked and serviced by a licensed electrician within the past year?

e. Is emergency lighting operative in simulated power failure?

f. Has periodic inspection and service of all elevator equipment been provided?

g. Are escalator emergency shut-off buttons operative; has periodic inspection and service of all escalator equipment been provided?

21. FIRE PROTECTION

a. Are adequate fire extinguishers in place, accessible, ready for use—tags dated and signed—*employees trained in use?*

b. Are sprinkler heads unobstructed?
(Minimum clearance 18",—36" when storage height exceeds 15'.)

c. Is fire hose in place, unobstructed and ready for use?

d. Are flammable wastes (wrappings, tissues, etc.) stored in covered metal containers and removed from the premises daily or more frequently if necessary?

e. Are flammable liquids, paints, thinners, etc., safely stored and used in a safe manner?

f. When installed is the fire alarm system under the supervision of qualified persons who test it weekly?

(1) Are all employees instructed in its use,
(2) Is the fire department telephone number properly posted?

g. Are adequate and strategically placed ash trays provided, floors free of cigarette butts and "No Smoking" signs posted in prohibited areas?

h. Are all automatic and self-closing fire doors operative? Fire doors to stairs kept closed? Moveable parts of fire escapes operable?

i. Where remodeling work, painting or decorating is in progress, is an inspection made at the end of each day to safeguard the area?

j. Are pressing irons always "parked" on proper stands and turned off during long idle periods and at the end of the working day?

22. MISCELLANEOUS—SAFETY

a. Do Employees know the procedure in event of an accident to a customer or fellow employee?

	Y	N
MISCELLANEOUS (cont'd)		
b. Are the necessary emergency telephone numbers correct and properly posted—police, fire, medical, security?		
c. Are employees knowledgeable in procedures for emergency evacuation of all store occupants?		
d. Are all hand tools (scissors, knives, hammers, etc.) in safe condition and properly stored? Power tools properly guarded?		
e. Is First Aid Kit adequate and kept properly filled?		
f. Is the OSHA poster conspicuously displayed?		
g. Do all electric ventilating fans within reach have a protective shroud with openings small enough to preclude accidental blade injury?		
h. Do all sewing machines have spring type needle guards and/or plastic eye shields?		

A Selected Bibliography

Anderson, Ronald A. *Wharton's Criminal Law and Procedure.* Rochester Lawyers Cooperative Publishing Co., 1957.

Astor, Saul D. *Loss Prevention: Controls and Concepts.* Los Angeles: Security World Publishing Co., Inc., 1978.

Barefoot, J. Kirk (Ed.) *The Polygraph Story.* Linthicum Heights, Md.: American Polygraph Association, Revised Third Printing, October, 1974.

Barefoot, J. Kirk. *Undercover Investigation.* Springfield, Ill.: Charles C. Thomas, 1975.

Caesar, Gene. *Incredible Detective.* Englewood Cliffs, N.J.: Prentice-Hall, Inc., 1968

Cevetic, Matthew. *The Big Decision.* Los Angeles: 1959.

Ellis, Bill. "Proof of Loss in Fidelity Bond Claims." *Security World* magazine, May, 1978.

Lee, C.D. *The Instrumental Detection of Deception.* Springfield, Ill.: Charles C. Thomas, 1953.

Leininger, Sheryl (Ed.) *Internal Theft: Investigation and Control.* Los Angeles: Security World Publishing Co., Inc., 1975.

Levy, Robert. "The Big Rip-Off in Purchasing." *Dun's Review,* March, 1977.

Moscarello, Grau and Chapman. *Retail Accounting and Financial Control.* Fourth Edition. New York, N.Y.: Ronald Press, 1976.

Motto, Carmine J. *Undercover.* Springfield, Ill.: Charles C. Thomas, 1971.

Reid, John E. and Inbau, Fred E. *Truth and Deception: The Polygraph Technique.* Second Edition. Baltimore, Md.: Williams & Wilkins Co., 1976.

Smith, Lawrence D. *Counterfeiting.* New York, N.Y.: Norton, 1944.

Walsh, Timothy J. and Healy, Richard J. *Protection of Assets.* Santa Monica, Ca.: The Merrill Co., 1974.

INDEX

Index

C

D

E

F

G

H

I

T

U

V

W

Other Security Books from Butterworths . . .

AIRPORT, AIRCRAFT & AIRLINE SECURITY/Moore

ALARM SYSTEMS AND THEFT PREVENTION/Weber

BANK SECURITY/Anderson

CAMPUS SECURITY/Powell

COMPUTER SECURITY/Carroll

CONFIDENTIAL INFORMATION/Carroll

EFFECTIVE SECURITY MANAGEMENT/Sennewald

EMPLOYEE THEFT INVESTIGATION/Barefoot

FORMS FOR SAFETY AND SECURITY MANAGEMENT/Guy, et al

THE HANDBOOK OF LOSS PREVENTION & CRIME PREVENTION/Fennelly

HOSPITAL SECURITY, 2nd Edition/Colling

HOTEL & MOTEL SECURITY MANAGEMENT/Buzby & Paine

INDUSTRIAL SECURITY/Berger

INTERNAL THEFT

INTRODUCTION TO SECURITY/Green

INTRUSION DETECTION SYSTEMS/Barnard

KINKS & HINTS FOR THE ALARM INSTALLER

LOSS PREVENTION/Astor

MANAGING EMPLOYEE HONESTY/Carson

MANAGING INFORMATION SECURITY/Schweitzer

OFFICE & OFFICE BUILDING SECURITY/SanLuis

THE PROCESS OF INVESTIGATION/Sennewald

RESTAURANT AND BAR SECURITY

SECURITY ADMINISTRATION/Morneau

SECURITY FOR SMALL BUSINESSES/Berger

SECURITY PROBLEMS IN A MODERN SOCIETY/Strauss

SECURITY SUPERVISION/Finneran

SUCCESSFUL RETAIL SECURITY

SUPERVISORY TECHNIQUES FOR THE SECURITY PROFESSIONAL/ Wanat, et al

UNDERSTANDING AND SERVICING ALARM SYSTEMS/Trimmer

VIDEO SECURITY/Bose